Hosting Earth

Hosting Earth is a timely and much-needed volume in the emerging literature of environmental philosophy, drawing upon art, science, and politics to explore alternatives to the traditional domination of nature by humans.

Featuring a dialogue with Mary Robinson (former UN High Commissioner for Human Rights and former President of Ireland), which addresses the current climate emergency, this book engages the question of ecological hospitality: what does it mean to be guests of the earth as well as hosts? It includes chapters by cutting-edge scholars in the philosophy of nature, as well as artists, scientists, psychologists, and theologians. The contributors discuss proposals for a new "Poetics of the Earth," opening horizons beyond our perilous Anthropocene to a new Symbiocene of mutual collaboration between human and non-human species.

Focusing on the central role that the human psyche plays in answering our current ecological emergency, *Hosting Earth* is for anybody invested in the future of our planet and how psychological, psychoanalytic, and philosophical thought can reorient the current conversation about ecology.

Richard Kearney is Charles Seelig Chair of Philosophy at Boston College and author and editor of more than forty books on contemporary philosophy and culture. He is founding director of The Guestbook Project for Narrative Hospitality and has been engaged in developing a postnationalist philosophy of peace and empathy over several decades.

Peter Klapes is a graduate in Philosophy at Boston College. His main philosophical interests include psychoanalysis, the philosophy of literature, and contemporary continental philosophy and his writing has appeared in a number of international journals. Peter currently serves as Executive Manager of The Guestbook Project.

Urwa Hameed studied Political Science, International Relations, and Managing for Social Impact at Boston College. She has written on the role of women in politics in Pakistan, and is also a founder and President of the non-profit, Free Immigrations Services.

The Psychology and the Other Book Series

Series editor: David M. Goodman
Associate editors: Matthew Clemente, Brian W. Becker,
Donna M. Orange and Eric R. Severson

The *Psychology and the Other* book series highlights creative work at the intersections between psychology and the vast array of disciplines relevant to the human psyche. The interdisciplinary focus of this series brings psychology into conversation with continental philosophy, psychoanalysis, religious studies, anthropology, sociology, and social/critical theory. The cross-fertilization of theory and practice, encompassing such a range of perspectives, encourages the exploration of alternative paradigms and newly articulated vocabularies that speak to human identity, freedom, and suffering. Thus, we are encouraged to reimagine our encounters with difference, our notions of the "other," and what constitutes therapeutic modalities.

The study and practices of mental health practitioners, psychoanalysts, and scholars in the humanities will be sharpened, enhanced, and illuminated by these vibrant conversations, representing pluralistic methods of inquiry, including those typically identified as psychoanalytic, humanistic, qualitative, phenomenological, or existential.

Recent titles in the series include:

Levinas for Psychologists
Leswin Laubscher

The Psychosis of Race
A Lacanian Approach to Racism and Racialization
Jack Black

Meaningless Suffering
Traumatic Marginalisation and Ethical Responsibility
Edited by David Goodman and M. Mookie C. Manalili

Hosting Earth
Facing the Climate Emergency
Edited by Richard Kearney, Peter Klapes and Urwa Hameed

For a full list of titles in the series, please visit the Routledge website at: https://www.routledge.com/Psychology-and-the-Other/book-series/PSYOTH

Hosting Earth

Facing the Climate Emergency

Edited by Richard Kearney,
Peter Klapes and Urwa Hameed

Routledge
Taylor & Francis Group

LONDON AND NEW YORK

Designed cover image: © Simone Kearney, "hands bloom," soapstone and marble, 4 1/2 x 10 1/2 x 10 1/2 in, 2023.

First published 2025
by Routledge
4 Park Square, Milton Park, Abingdon, Oxon OX14 4RN

and by Routledge
605 Third Avenue, New York, NY 10158

Routledge is an imprint of the Taylor & Francis Group, an informa business

British Library Cataloguing-in-Publication Data
A catalogue record for this book is available from the British Library

ISBN: 978-1-032-59948-9 (hbk)
ISBN: 978-1-032-59949-6 (pbk)
ISBN: 978-1-003-45694-0 (ebk)

DOI: 10.4324/9781003456940

Typeset in Times New Roman
by KnowledgeWorks Global Ltd.

In memory of James Mahoney and Brian O'Donovan—
generous guiding spirits of the Guestbook Project

Contents

Contributors

Kate Burrows, MPH, PhD is an Assistant Professor in the Department of Public Health Sciences at the University of Chicago. Dr. Burrows's research is primarily focused on the impacts of climate change on public health. She has studied a wide range of environmental extremes including heatwaves, flooding, landslides, and hurricanes, and their effects on both physical and mental health outcomes. Dr. Burrows employs an interdisciplinary approach that integrates sociocultural determinants of health and environmental factors in her work. She utilizes a combination of research methods, conducting qualitative and community-based participatory studies at the local level and quantitative analyses using national-level datasets. Before joining the University of Chicago, Dr. Burrows served as a Voss Postdoctoral Fellow at the Institute at Brown University for Environment and Society. She holds a PhD in Environmental Epidemiology from Yale University School of the Environment and an MPH in Social-Behavioral Sciences from Columbia University's Mailman School of Public Health.

Edward Casey is an American philosopher who has published ten volumes on phenomenology, philosophical psychology, and the philosophy of space and place. His work is widely cited in contemporary continental philosophy. He was elected President of the American Philosophical Association (Eastern division) in 2010. He founded the MA in Philosophy and the Arts located in Brooklyn. He is currently Distinguished Professor of Philosophy Emeritus at Stony Brook University in New York.

Brian R. Clack is Professor of Philosophy and A. Vassiliadis Director of the Humanities Center at the University of San Diego. He is the author of two books on the philosophy of Wittgenstein—*Wittgenstein, Frazer and Religion* (1999) and *An Introduction to Wittgenstein's Philosophy of Religion* (1999)—and of *Love, Drugs, Art, Religion: The Pains and Consolations of Existence* (2014). He is co-author, with Beverley Clack, of *The Philosophy of Religion: A Critical Introduction*, now in its third edition (2019). With Tyler Hower, he edited the anthology *Philosophy and the Human Condition* (2018). Most recently Clack has edited works by Edmund Burke and the Marquis of Halifax.

Matthew Clemente is a husband and father. He is a Research Fellow in the Center for Psychological Humanities and Ethics at Boston College and the author of several books, most recently *Posttraumatic Joy: A Seminar on Nietzsche's Tragicomic Philosophy of Life*.

Melissa Fitzpatrick is an Assistant Professor of Business Ethics and Sustainability in the College of Business Administration at Loyola Marymount University, Los Angeles. Chief among her research interests is understanding how to foster a more sustainable community within and beyond the human world, and, as a vital foundation for that, how to overcome instrumental values. She earned her PhD in philosophy from Boston College, an MA in philosophy from Loyola Marymount University, and a BS in communication from Boston University. She is the co-author of *Radical Hospitality: From Thought to Action*.

Matthias Fritsch is Professor of Philosophy at Concordia University. His research interests are social and political philosophy, environmental ethics, and 19th- and 20th-century European philosophy (especially German critical theory and deconstruction). He is the author, among other books, of *Taking Turns with the Earth: Phenomenology, Deconstruction, and Intergenerational Justice*.

Lauren Guilmette is an Associate Professor of Philosophy with affiliations in Women's, Gender, and Sexualities Studies and Environmental Studies at Elon University. Her research engages theories of affect and feeling, particularly the work of the late feminist theorist Teresa Brennan. Her articles have appeared in journals such as *philoSOPHIA*, *differences*, and *The Journal of Speculative Philosophy*.

Marion Hourdequin is Professor of Philosophy at Colorado College, and specializes in environmental philosophy. She is the author of *Environmental Ethics: From Theory to Practice* and editor, with David Havlick, of *Restoring Layered Landscapes*. Professor Hourdequin is Vice President / President Elect of the International Society for Environmental Ethics and serves as an Associate Editor for two journals, *Environmental Values* and *Environmental Ethics*.

Fanny Howe is an American poet, novelist, and short story writer. Howe has written more than twenty books of poetry and prose. Her major works include poetry such as *One Crossed Out*, *Gone*, and *Second Childhood*, the novels *Nod*, *The Deep North*, and *Indivisible*, and collected essays *The Wedding Dress: Meditations on Word and Life* and *The Winter Sun: Notes on a Vocation*. She was awarded the 2009 Ruth Lilly Poetry Prize by the Poetry Foundation and is also the recipient of the Gold Medal for Poetry from the Commonwealth Club of California.

Leah Kalmanson is an Associate Professor and the Bhagwan Adinath Professor of Jain Studies at the University of North Texas. She is the author of the 2020 book *Cross-Cultural Existentialism* and co-author of the 2021 *A Practical Guide to World Philosophies*. Her essays appear in journals including *Comparative* *and Continental Philosophy, Continental Philosophy Review, Frontiers of Philosophy in China, Hypatia, Journal of World Philosophies, Philosophy East & West, Pragmatism Today, Shofar,* and *Studies in Chinese Religions*, as well as the digital magazine *Aeon.*

Michael Kearney, MD, has spent over thirty years working as a physician in end-of-life care. He trained and worked at St. Christopher's Hospice in London with Dame Cicely Saunders, the founder of the modern hospice movement, and subsequently worked for many years as Medical Director of Our Lady's Hospice in Dublin, and later with Professor Balfour Mount at McGill University in Montreal. He is currently a Medical Director of the Palliative Care Consultation Service at Santa Barbara Cottage Hospital and an Associate Medical Director at Visiting Nurse and Hospice Care, also in Santa Barbara. Dr. Kearney teaches nationally and internationally and is the author of three published books, including *Mortally Wounded: Stories of Soul Pain, Death, and Healing,* and *A Place of Healing: Working with Nature and Soul at the End of Life,* as well as many articles and chapters. His most recent book, *The Nest in the Stream: Lessons from Nature on Being with Pain,* offers a way of being with pain and suffering that is infused with mindfulness, nature connection, openness, and compassion.

Catherine Keller is George T. Cobb Professor of Constructive Theology in the Graduate Division of Religion of Drew University. Her work falls along a spectrum of pluralist, process, and political philosophy along with ecofeminist religious and theological variants. Books she has authored include *From a Broken Web: Separation, Sexism and Self; Apocalypse Now & Then: A Feminist Guide to the End of the World; God & Power; Face of the Deep: A Theology of Becoming; On the Mystery: Discerning God in Process; Cloud of the Impossible: Negative Theology and Planetary Entanglement; Intercarnations: Exercises in Theological Possibility; Political Theology of the Earth: Our Planetary Emergency and the Struggle for a New Public.* Most recent is *Facing Apocalypse: Climate, Democracy and Other Last Chances.* She has co-edited several volumes of the *Drew Transdisciplinary Theological Colloquium,* including *Postcolonial Theologies; Ecospirit; Apophatic Bodies; Polydoxy; Common Good/s: Ecology, Economy and Political Theology;* and *Entangled Worlds: Religion, Science and the New Materialism.*

Michael Robert Kelly is Professor of Philosophy at the University of San Diego. He is author of *Phenomenology and the Problem of Time* (2016) and *A Phenomenological Analysis of Envy* (2024), editor of *Bergson and Phenomenology* (2010), and co-editor of *Michel Henry: The Affects of Thought* (2012), *Early Phenomenology* (2016), and *Michel Henry's Practical Philosophy* (2022).

John P. Manderson, PhD, OpenOcean Research, specializes in developing collaborative industry–science partnerships and research projects to improve the accuracy and precision of marine resource assessments used to inform harvest regulations. Before founding OpenOcean Research, Dr. Manderson worked for 23 years at the US National Oceanographic and Atmospheric Administration as a marine ecosystem scientist, research fisheries oceanographer, and as an expert in industry–science collaborative research. While at NOAA, he successfully formed and led trans-disciplinary groups of experts from the fishing industry, government, academia, and environmental NGOs focused on integrating ecosystem considerations including climate change into assessments of fish populations.

Jane D. Marsching is an interdisciplinary artist who explores our past, present, and future human impact on the environment through collaborative research-based practices. Projects have been sited in museums and galleries as well as weather observatories, public parks, city streets, radio waves, and the internet. She has worked with scientists, educators, kite builders, meteorologists, architects, and musicians, among others. Recent exhibitions include: Sun Valley Center for the Arts, Ketchum, Idaho; University of Massachusetts, Boston; Northern Spark, Minneapolis, MN; Kilroy Metal Ceiling, Brooklyn; Galerie Lucy Mackintosh, Lausanne, Switzerland; Tierra des Explorades, Buenos Aires; and MassMoCA. She has received grants from Creative Capital and Artadia, among others. In another time, she was a book editor, most recently coediting a book of essays by 12 authors entitled *Far Field: Digital Culture, Climate Change, and the Poles* with Andrea Polli in 2012, and *Blur of the Otherworldly: Contemporary Art, Technology, and Paranormal* with Mark Alice Durant in 2006. She occasionally writes about ecology and art; a recent essay was included in *Running Falling Flying Floating Crawling*, edited by Mark Alice Durant and published by Saint Lucy Press in 2021. At Massachusetts College of Art and Design, she is Professor and Director of the Minor in Sustainability.

Sean McGrath has published widely in the history of ideas and the philosophy of religion. He is Full Professor of Philosophy at Memorial University and Adjunct Professor of Religious Studies at McGill University. He is the author of *Thinking Nature: An Essay in Negative Ecology* (2019) and *Political Eschatology* (2023).

James Morley is a Professor of Clinical Psychology at Ramapo College of New Jersey where he teaches Psychopathology, Phenomenological Psychology, and Social Theory. He is the Editor-in-Chief of the *Journal of Phenomenological Psychology*, Founding Director of Ramapo's Mindfulness Center and is the recent Past President of the Interdisciplinary Coalition of North American Phenomenologists. Morley's publications and research interests are in psychology as a human science and the application of phenomenological thought to qualitative methodology and the topics of imagination, mental health, and Asian thought.

Marjolein Oele is Professor of Philosophy of the Humanities at Radboud University (the Netherlands), and Professor of Philosophy at the University of San Francisco. Her research intertwines Ancient Philosophy, Continental Philosophy, and Environmental Philosophy. She is the author of *E-Co-Affectivity: Exploring Pathos at Life's Material Interfaces* (2020) and co-editor of *Ontologies* *of Nature: Continental Perspectives and Environmental Reorientations* (2017). Her most recent book project is entitled *Elemental Loss: Shifting Constellations of Water, Fire, Air and Earth.* Her articles have been published in a range of journals, including *Ancient Philosophy, Configurations, Environmental Philosophy, Epochê, Radical Philosophy Review,* and *Research in Phenomenology.* She is the Editor-in-Chief of the journal *Environmental Philosophy.*

Joseph S. O'Leary (born, Cork, 1949) studied literature and theology in Maynooth, Rome, and Paris, taught theology in the USA, taught literature in Sophia University, Tokyo, and did research on Buddhism in Nanzan University, Nagoya. His publications include *Conventional and Ultimate Truth: A Key for Fundamental Theology* (2015), *Joysis Crisis* (2021) and *Irreducible Ireland* (2024).

Donna M. Orange, PhD, PsyD, educated in philosophy, clinical psychology, and psychoanalysis, teaches at NYU Postdoc (New York University Postdoctoral Program in Psychoanalysis and Psychotherapy) and at IPSS (Institute for the Psychoanalytic Study of Subjectivity, New York), and in private study groups. She held the Freud Fulbright in Vienna in 2018. Recent books are *Thinking for Clinicians: Philosophical Resources for Contemporary Psychoanalysis and the Humanistic Psychotherapies* (2010), *The Suffering Stranger: Hermeneutics for Everyday Clinical Practice* (2011), *Nourishing the Inner Life of Clinicians and Humanitarians: The Ethical Turn in Psychoanalysis* (2016), *Climate Justice, Psychoanalysis, and Radical Ethics* (2017), and *Psychoanalysis, History, and Radical Ethics: Learning to Hear* (2020).

Graham Parkes taught Asian and comparative philosophy in the Philosophy Department at the University of Hawai'i at Mānoa for almost 30 years. In 2008 Parkes moved to University College Cork in Ireland, where he was the Professor of Philosophy and then Head of the School of Philosophy and Sociology. He has held research appointments in France, Italy, and Japan, and has been a Visiting Professor in Austria, China, Japan, and Singapore. He is currently a Professorial Research Fellow in Philosophy at the University of Vienna. His most recent book is *How to Think About the Climate Crisis.*

Chandler D. Rogers is Lecturer in Philosophy at Gonzaga University in Spokane, Washington. He is interested in thinking about the human place in nature within the modern cosmological context, in relation to divinity and to creatures, and his work has been published in journals such as *International Journal for Philosophy of Religion, Environmental Philosophy, Journal for Critical Animal Studies, Philosophy and Theology,* and *Worldviews: Global Religions, Culture, and Ecology.*

Ariel Salleh is Distinguished Visiting Scholar in the Centre on Labour, Sustainability and Global Production, Queen Mary University of London in 2023 and Visiting Professor, Faculty of Humanities, Nelson Mandela University, South Africa. Dr. Salleh's transdisciplinary writing is seminal to political ecology as an emerging study of humanity-nature relations. Her 'embodied materialist' approach emphasizes the centrality of reproductive or regenerative labour in the world system.

David Storey is Associate Professor of the Practice in the Philosophy Department at Boston College, where he teaches courses in the Great Books, environmental ethics, and philosophy of technology. He is the author of *Naturalizing Heidegger* (2015); hosts the podcast Wisdom at Work: Philosophy Beyond the Ivory Tower, which features philosophers working outside the academy; and is a Senior Fellow at the Institute for Cultural Evolution.

Olúfẹ́mi O. Táíwò is Associate Professor of Philosophy at Georgetown University. His public philosophy, including articles exploring intersections of climate justice and colonialism, has been featured in *The New Yorker*, *The Nation*, *Boston Review*, *Dissent*, *The Appeal*, *Slate*, *Al Jazeera*, *The New Republic*, *Aeon*, and *Foreign Policy*. He is the author of *Elite Capture* and *Reconsidering Reparations*.

Brian Treanor is Professor of Philosophy at Loyola Marymount University, where he teaches courses in environmental philosophy, philosophy and literature, and philosophy of religion, among other subjects. In 2011, he was awarded the President's Fritz B. Burns Distinguished Teaching Award. He is the author or co-editor of ten books, including: *Melancholic Joy* (2021), *Philosophy in the American West* (2020), *Carnal Hermeneutics* (2015), *Being-in-Creation* (2015), *Emplotting Virtue* (2014), and *Interpreting Nature* (2013). He is currently working on two monographs, one exploring the meaning of "nature" and "wilderness," and the other arguing for the selfhood of non-human nature.

Stanley Uche Anozie is Assistant Professor of the Practice in the Philosophy Department at Boston College, where he teaches courses on ethics, social and political thought, and philosophy of race. He is an editorial board member of *International Journal of Philosophy*.

Lisa and Wolf Wahpepah are native elders who run Descendants of the Earth, a community-based nonprofit organization reestablishing an Inter-Tribal Spirit Camp in the high desert of Central California. They are the second generation of their family to carry a traditional Native American Fireplace for the benefit of all people, promoting the values of Native American culture, including harmony in interpersonal relations, peaceful conflict resolution, and environmental restoration of Mother Earth.

Rowan Williams is a Welsh Anglican bishop, theologian and poet. He was the 104th Archbishop of Canterbury, the 35th Master of Magdalene College, and is a Honorary Professor of Contemporary Christian Thought in the University of Cambridge. He is the author of many books, most recently Passions of the Soul, Looking East in Winter: Contemporary Thought and the Eastern Christian Tradition, and The Way of St Benedict.

Jason M. Wirth is Professor and Chair of Philosophy at Seattle University and works and teaches in the areas of Continental Philosophy, Buddhist Philosophy, Aesthetics (especially film, painting, poetry, and the novel), and Environmental

Philosophy. His recent books include *Nietzsche and Other Buddhas: Philosophy after Comparative Philosophy* (2019) and *Mountains, Rivers, and the Great Earth: Reading Gary Snyder and Dōgen in an Age of Ecological Crisis* (2017). He is the Associate Editor and Book Review Editor of the journal *Comparative and Continental Philosophy*, and is currently completing a manuscript on the cinema of Terrence Malick as well as a work of ecological philosophy called *Turtle Island Anarchy*.

Christopher Yates teaches philosophy at James Madison University in Harrisonburg, Virginia, and is a Research Fellow at the Institute for Advanced Studies in Culture, the University of Virginia. He specializes in 19th–20th century European Philosophy, and Aesthetics, and is the author of *The Poetic Imagination in Heidegger and Schelling* (2013). His work has appeared in journals such as *Research in Phenomenology*, *Comparative and Continental Philosophy*, *Continental Philosophy Review*, *The European Journal of Philosophy*, and *The Hedgehog Review*.

Stefano Zamagni (born Rimini, 1943), Professor of Economics at the University of Bologna, is Emeritus President of the Pontifical Academy of Social Sciences and a fellow of the Human Development and Capability Association. His publications include *L'economia del bene comune* (*The Economics of Common Good*) (2008) and *An Outline of the History of Economic Thought* (co-authored, 1995 and 2005).

Michael E. Zimmerman was Professor of Philosophy at Tulane University for 31 years, then taught at the University of Colorado Boulder, retiring as Emeritus Professor of Philosophy in 2015. He published widely in environmental philosophy, integral theory, Heidegger and Nietzsche, and Buddhism. His last book was *Integral Ecology: Uniting Multiple Perspectives on the Natural World* (with Sean Esbjörn Hargens), and he is also the author of a widely used anthology on environmental philosophy, *Environmental Philosophy: From Animal Rights to Radical Ecology*.

Acknowledgments

Hosting Earth endeavors to give voice to new conversations in environmental philosophy. It engages the vital question of ecological hospitality: what does it mean to be guests of the earth as well as hosts? The book has its origins in a conference organized by the Guestbook Project at Boston College in the Spring of 2022. It opens with a dialogue with world policy maker Mary Robinson (former Special UN Envoy on Climate Justice and Commissioner of Human Rights) exploring ways of addressing the current climate emergency; and includes contributions by cutting-edge scholars in the philosophy of nature, as well as artists, life scientists, psychologists and theologians. One of the key proposals is a new "poetics of earth," pointing beyond our imperiled Anthropocene to a Symbiocene of mutual collaboration between humans and non-humans. A signal feature of the volume is its radically interdisciplinary nature, fostering a vibrant colloquy between philosophy, art, science, and politics.

We are grateful to a number of people and organizations for making this publication possible, most notably the Guestbook Project in Narrative Hospitality, Write the World, the Irish-American-Partnership, and the Center for Psychological Humanities and Ethics at Boston College. We would also like to thank the following for their generous and collegial collaboration and assistance—David Goodman, Sheila Gallagher, Matt Clemente, David Storey, Aidan Browne, Laura Steinberg, Guy Beiner, Robert Savage, Dermot Moran, Anne Bernard Kearney, Mary Sugrue, Maria Gallego Ortiz, Magnus Ferguson, Jaret Farhat, Alex Riedel, Jeronimo Ayesta, and Jared Highlen. And, finally, a word of deep gratitude to Zoe Meyer, Jana Craddock, Victoria Leigh, Tom Bedford, and all the editorial team at Routledge.

Preface: Towards a Hospitality of Nature

Richard Kearney

We are born of the earth and return to it in the end. As the earliest myths remind us, humans were made from mud—*humus*—and it is with humble humanity that we recall our terrestrial origins. One of our primary callings as humans is to remain faithful to the earth which bore us, and the call will not go silent for as long as we dwell on this planet.

Today the earth cries out. Our ecological crisis has become an emergency. As several of the contributors to this volume note, we have been aware of a growing climate crisis for almost half a century. Scientists have issued warnings for decades now. But we are still not making the critical changes necessary to save existing life on this planet. The COP28 was another chilling reminder that we are moving perilously toward the 4-degree mark. And even if we manage to reduce carbon emissions to net 0 by 2050, we are still likely to face a climatological legacy of perilous proportions. Wildfires, storm floods, melting glaciers, heatwaves, and shore erasures are on the increase. We are faced with an alarming biodiversity collapse with the global wildlife population shrinking by over 60 percent in the last fifty years. We have polluted the earth upsetting the chemical composition of its atmosphere and the pH of its oceans. Microplastics contaminate our soil and sea, falling in rain from the sky and circulating throughout the body's blood, even in placentas and breast milk. There is no place immune to the threat of human appropriation.[1] The scientific data is irrefutable; and there is hardly a school child today unable to explain Global Warming or the Greenhouse Effect. A recent UNESCO questionnaire showed that our climate emergency is now recognized worldwide as our most pressing ethical concern, along with poverty and migration which are inextricably linked. We are suffering from a severe Nature Dissociation Disorder.[2] And yet, with few exceptions, the political will to take action seems unconscionably absent. So what is to be done?

We might begin with a new way of thinking. A paradigm shift is urgently needed in our vision of the earth, seeing it as a host that nurtures us rather than an object to be exploited. For real change to occur, we must heed the cry of the earth and embrace a radical ecological hospitality toward all living beings. Such a shift would require us to move from the Anthropocene of relentless human domination—which has brought us to the brink of disaster—to a Symbiocene of deep interdependency between humans and nature. It is surely time for the *Anthropos*, defined by Plato as the being who looks up and away from nature, to rediscover its primal belonging as *symbiosis*: living together with different species in mutual dependency.[3]

But what might this entail? This volume hopes to offer some ideas by inviting a number of environmental philosophers, scientists, and artists to address this urgent task. In the face of imminent climate collapse, we can still imagine possibilities of resilient resistance and reorientation. As Mary Robinson says in our opening conversation, we must not succumb to despair but embrace a number of basic tasks. First, we have the *moral* task of making the climate crisis count in our everyday lives (e.g. responsible production and consumption: recycle, reuse, and refuse pollutant energies). Second, we have the *political* response of righteous indignation towards those in positions of power who are not doing what they should —public advocacy and protest. And third, we have the *practical* duty to support NGOs, multilateral civil societies and other organizations creating grassroots green teams working for change. The last of these reminds us that we are all in this together. When it comes to sharing our endangered environment, we are irrevocably interconnected, whether we like it or not. And this calls for a fourth step—a *pedagogical* commitment to urgent philosophical conversations about what needs to be done. An act of "conscientization" regarding the mutual interdependency between humans and non-humans. This fourth task marks a transition to a Symbiocene of radical interspecies hospitality, based on practices of environmental justice and solidarity. A move that requires that we cease being victims of apocalypse and become instead "prisoners of hope."

In the beginning was symbiosis; and the earth can still be our teacher if we acknowledge it as host and ourselves as guests. Trees, for example, are a perennial example of how symbiosis operates. They flourish and communicate by using subterranean fungal networks, which channel the flow of life resources and information throughout forests. This phenomenon of interaction is confirmed by the research of Suzanne Simard and other environmental scientists showing how forests have "mother trees" or large interconnecting hubs whose underground mycorrhizal relationships contribute to forest resiliency, adaptability, and survival.[4] This has major implications for how to manage and heal forests from human harm, most notably the climate damage brought about by the worst excesses of the Anthropocene—to wit, modernity's determination to exploit nature, reducing living things to commodities of consumption and exchange. Our voluntarist anthropocentrism was epitomized by Descartes' boast that "man is master and possessor of nature"; and it finds symptomatic expression in the rapacious food, fashion, and fossil industries of our industrial market economies.

In contrast to this top-down hegemonic model, consider more closely the circular mutuality of the symbiotic alternative. At the most infinitesimal level of nature, mycorrhizal fungi live in reciprocal relationship with plants. The fungi serve as root extensions which transmit water and other basic nutrients for survival, while plants furnish the fungi with sugars in return. Beeches and birches, for instance, could not communicate without such complex mycorrhizal interactions. Indeed, recent scientific data reveals that 90 percent of all known plants grow in association with fungi, a symbiotic practice of hospitality going back 400 million years.

This mutual interdependency signals an underground matrix of collaboration, commonly known as the Wood Wide Web. And fully understanding how these terrestrial ecosystems operate—with fungi serving as critical decomposers and recomposers of life matter—is crucial to appreciating the importance of biodiversity for the survival of our planet. Without it the world's continental landmasses would become devoid of forests, crops, and grasslands. There would be no sustenance for living creatures, human and non-human alike.[5]

The cover image of this volume features the tree motif as *axis mundi*—a sheltering host of nature. It echoes Rilke's poetic hypothesis that "if we surrendered to earth's intelligence we could rise up rooted, like trees." In multiple wisdom traditions, trees embody a medial space between land and sky, considered a middle world connecting the earth beneath our feet and the air above our heads. The Buddha, for example, achieved enlightenment by breathing in harmony with a Bodhi tree. The Chinese Cosmic Man, P'an Ku, was shrouded in arboreal leaves. Christ enacted the death–life cycle by offering himself to the world on a tree grown from the root of Jesse.[6] Indeed, one of the seminal motifs of all world religions is the "Great mother and her symbol, the Tree," epitomized by a certain indigenous belief in the "bush soul" that becomes incarnate in living plants, which in turn host and protect humans as guests.[7] This panpsychist belief is still held by many indigenous people today, as the Native American elder, Lisa Wahpepah, reminds us in this volume. When one thinks of the Earth like a mother, she writes, living creatures are seen as life givers. "They nurture us and give us everything to live, not just physically, but also mentally and emotionally. That is why all living beings are our relations, not only the two-legged but the four-legged, the winged, the insects, the finned that live in the ocean, the trees, the plankton, the flowers. It is everything, all my relations. *Mitakuye Oyasin.* When we understand our intimate relationship with the Earth we are in touch with what the Earth gives to us."[8] But few of us live in that symbiotic relationship today, as climate clouds darken our skies. We have become "disconnected" from our elemental relation to the earth, no longer acknowledging her as mother. Lisa's fellow elder, Wolf Wahpepah, adds this moving and candid caution: "The natural world, the Earth, regards all life and wants to give life. It is not a destroyer. It is a creator. We're the ones who have thrown the life cycle off balance to such an extent that now we threaten the organism that sustains all life. At some point in time Mother Earth has to decide how much can be salvaged because she has to make sure that life itself is not extinguished. If she has to make decisions or cause consequences that make it less habitable for us in order to diminish our negative impact… she will. Any mother would. And that is why it is better to think of her as a mother as opposed to objectifying her. The earth is not *like* our mother, It *is* our Mother. And how you care for your mother is how you should care for the earth."[9]

Another vocal exponent of ecological hospitality is the indigenous scientist, Robin Kimmerer, a nature writer of the Potawatomi nation and Professor of Environmental Biology at the State University of New York. In her influential book *Braiding Sweetgrass* (2013), she celebrates the indigenous agricultural practice of the Three Sisters: namely, three alimentary seeds which grow in symbiotic harmony. "Together these plants—corn, beans and squash—feed the land, and feed our imaginations, telling us how we might live" (p. 127). Indeed the Potawatomi language construes most of the world as a composite interaction between species, using verbs rather than nouns. The mountain is not a fixed object amongst objects but a way of becoming a mountain. A "mountaining" in eco-communion with neighboring ways of "treeing" and "skying" and "rivering" and "wolfing." Only things fabricated by human subjects are called objects—the rest are animate beings deserving reverence and respect.[10] Which is not to deny that nature can be "naturing" in multiple ways. While nourishing us, it also harbors shadows and sufferings, decomposition as well as growth, death drives as well as life drives, all convening in a complex interplay of forces.[11]

The interspecies paradigm espoused by such indigenous thinking does not mean a regressive return to the past. On the contrary, it is very much in keeping with contemporary research in the sciences. It is no accident that nature writers like Kimmerer and Simard are both professional scientists and animists. Moreover, their writings about the work of symbiosis in nature chime felicitously with some of the keenest observations of astrophysics. What works on the ground is echoed in the skies. Take, for example, the phenomenon of cosmic eco-hospitality at work in the earth's upper atmosphere or ionosphere which acts like the membrane of a cell, keeping harmful matter and excess radiation out and allowing warmth and light in. The earth's magnetic envelope is analogous to a breathing in and out of energies while the "habitable zones" of planets operate in terms of an interdependency of external conditions (distance from star, mass of planet, presence of atmosphere) and internal conditions (preservation of habitats suitable for life). Similarly our solar system itself involves actions of dynamic equilibrium which take the form of cosmic interdependency. And what is true of astrophysics is equally true of the life sciences which attest to the existence of multiple ecosystems of reciprocal interaction, known as "symbiogenesis."[12]

It is curious how many contemporary scientists of matter have sought out conversations with some of the world's great spiritual leaders—think of David Bohm's exchanges with Krishnamurti, the Dalai Lama's dialogues with Harvard neurologists, Pope Francis' engagement with environmental scientists in *Laudato si'*. Indeed several contributors to this volume suggest that the four steps of ecological hospitality—personal, moral, political and pedagogical—may invite a fifth "spiritual" dimension. This persuasion is represented here, for example, by Buddhist thinkers like Jason Wirth and Michael Kearney (inspired by the teachings of Joanna Macy and Thich Nhat Hahn), by the conversations on "Asian" spiritualities of nature moderated by David Storey, as well as by the essays of Christian theologians like Catherine Keller, Rowan Williams, and Joseph O'Leary, who commends ecological poetics articulated by Hölderlin and Rilke. Indeed, it is true to say that almost every spiritual wisdom tradition harbors a special place for theophanies of the earth.[13]

One does not have to look as far as astrophysics or comparative theology, however, for fundamental paradigms of symbiotic existence. It is to be found in the act of breathing itself. We inhale oxygen produced by leaves which receive the carbon dioxide exhaled from our lungs. Every rising and falling breath participates in an exchange between our body and plants, mutually releasing and receiving chemical properties crucial for our respective flourishing. Each inhalation and exhalation embodies a reciprocal flow—simple, involuntary, ubiquitous, the most basic act of life from birth to death. Human hemoglobin and chlorophyll molecules in leaves are virtually identical, except that in hemoglobin the carbon rings encircle a molecule of iron, giving blood its red color, while in chlorophyll the very same ring structure encircles a molecule of magnesium giving leaves their greenness.[14] This act of bio-chemical mirroring constitutes a recursive dance of identity (ring structure) and difference (color), epitomizing a primal ecological collaboration between human and non-human life. "The world comes to breathe within me," as Bachelard writes. "Everything breathes in the world."[15]

The era of the Anthropocene has witnessed a massive deterioration of our terrestrial ecosystem. The facts are stark and incontrovertible. Almost half of the earth's surface has been impacted by human activity. The concentration of carbon in the earth's atmosphere has been increased by 30 percent since the onset of the industrial revolution, while recent history has seen the disappearance of over a quarter of the globe's bird species, with two thirds of the world's fish species now endangered or over-exploited.[16] But we must be wary of apocalyptic thinking. Too much environmental talk today is stuck in the horror-mode of "disgust, guilt and shame."[17] Alarmism is no substitute for action.

Let us return then to things of the earth. We often speak of nature as something outside of ourselves, but we are in fact deeply embedded in it from first to last. Humans are a natural incarnate species amongst others. For too long we have viewed nature as something there *for us*—a thing to be used, calculated, and consumed. We have treated it as a repository of experimental objects or as an amusement park for fantasies—a rustic wilderness which we romanticized or feared. It would be wiser, however, while there is still time, to see the earth as a living whole which hosts the human within it.[18]

The global market system has turned nature into a conduit of consumer goods, transforming plants and animals into food commodities for profit and exchange. There is no denying that modernity has brought extraordinary technological advances. It has given us many goods; but they are not all good. We need to discern between the deployment of technology to threaten our natural environment or to collaborate with the call of the earth in a new wager of ecological hospitality.[19] The choice is stark and challenging, but not impossible. And in choosing we may act as "edge dwellers," navigating between the twin demands of ecology and economy, technology and wilding, nature and culture.[20] Above all, we must recover our vocation as hosts and guests of the earth—sensitive and sensible partners in the breathing in and out of Being. And this is where we—*homo sapiens*—must return to the humor and humility of being thankful to the earth: the *humus* of our humanity which connects us to all living creatures, human and non-human alike. The very heart of nature herself.

> There really is inspiration and expiration of being, action and passion so slightly discernible that it becomes impossible to distinguish between what sees and what is seen... There is no break at all in this circuit and we are unable to say that nature ends here and human expression starts there. It is mute Being which itself comes to show forth its meaning.
>
> Maurice Merleau-Ponty.[21]

Notes

1 For these and further details about our climate emergency see the report of COP28 (28th United Nations Climate Change Conference 2023) as well as the contributions to this volume by Brian Treanor, Sean McGrath, David Storey, and Mary Robinson. (See especially Robinson, *Climate Justice: Hope, Resilience and the Fight for a Sustainable Future*, Bloomsbury, 2018). Shorter drafts of this Preface are available in the following journals: *Analecta Hermeneutica*, Vol. 14, No. 2, 2022, and *Symposium*, Vol. 29, Nos. 1–2, 2025.

2 See Richard Louv's ecological analysis of our "nature deficit disorder" in the Anthropocene. This deficit is symptomatic of a pathological dissociation which expresses itself as an acute "species loneliness" and, by reaction, as a "touch hunger" to return to our fundamental being-with-nature, in accordance with what Louv calls the "reciprocity principle" of mutual interspecies belonging (*The Wild Calling: How Connecting with Animals Can Save Our Lives and Save Theirs*, Algonquin, 2019). See our discussion of this theme in Richard Kearney, *Touch*, Columbia UP, 2021, pp. 109–111.

3 The Greek term *symbiosis* translates into Latin as *convivium*, "living together," one meaning of which is "feast." As Catherine Keller remarks: "The primal eucharist of life seems to find its genesis less in competition than in collaboration" (*The Cloud of the Impossible: Negative Theology and Planetary Entanglement*, Columbia UP, 2015, p. 157). Drawing on the research of biologist Lynne Margulis and mathematician Alfred North Whitehead, amongst others, Keller discusses the function of a wide range of "scientific symbioses" demonstrating the "elemental relationalism of microorganisms." Margulis speaks of an "interactive tissue of microbacteria converting the planet in a heretofore unknown layer of symbiosis" and possibly "explaining the evolution of life as an original, cooperative act of mutually constitutive relation called *symbiogenesis*." This refers to a relation of feeding: "but the imbibing of a single cell by another did not kill the first but enfolded it in a new creation—and so gave rise to complexity: the organism." Symbiogenesis is a primordial activity of collaboration in life, from the microscale of tiny organisms to the macroscale of the universe. Acknowledging this primal phenomenon of symbiosis may encourage us to replace the anthropocentric dualism of separate substances (e.g., *res cogitans* and *res extensa*) with a dynamic process of interactive becoming and belonging. So that particles, molecules, and cells—and each individual animal or plant composed of them—can be read as an "actual occasion constituted of its relations to all its others… a contraction of its universe from a unique perspective: it enfolds its universe and unfolds it differently" (Keller, pp. 175, 176–177). Symbiogenesis may thus be said to convene the opposites of identity and difference, unity and multiplicity—the physical universe mirroring the metaphysical interplay of the one and the many.

4 See Suzanne Simard, *Finding the Mother Tree: Discovering the Wisdom of the Forest*, Penguin Random House, 2021; and the very influential novel by Richard Power, *The Overrstory*, 2018, inspired by her work. These stories of arboreal symbiosis and reciprocity are echoed in the old childhood ditty, "I see the trees and the trees see me." See "How the Trees See Us," Guestbook Project Film 2021 on "Hosting Earth" Guestbookproject.org. This experience of a bilateral mirroring between humans and nature is powerfully captured in Merleau-Ponty's description of the reversible dialectic between seer and seen in "Eye and Mind" (*The Primacy of Perception*, Northwestern UP, 1964, p. 167. See also note 20 below).

5 See the review essay by Elizabeth Kolbert of Allison Pouliot's *Meetings with Remarkable Mushrooms: Forays with Fungi across Hemispheres*, U Chicago Press, 2023, in *The New York Review*, Vol. LXX, No. 14, Sept. 21, 2023, pp. 42–43. See also the important research by Michael Pollan on what might be called psychedelic symbiosis, *How to Change Your Mind*, Penguin Random House, 2018.

6 See the rood-tree of Christ blazing with animals, birds, and plants, sprouting from the root of Jesse in the altar crucifix hanging in the Basilica of San Clemente in Rome. See also Michael Kearney on the Buddha's enlightenment while breathing under the bodhi

tree, *Becoming Forest*, Eastern Studio Press, 2023; Mircea Eliade's treatment of the tree as a mythological and ritual *axis mundi* in *Patterns in Comparative Religion*; and Gaston Bachelard's reflections on the tree as an archetype of integration between the worlds of earth and sky (*On Poetic Imagination and Reverie*, Spring Publications, 2005 p. 84f).

7 Carl Jung, *Man and His Symbols*, Dell, 1968, pp. 6–7 and 69. This involved what Jung called a "mystical participation" or "unconscious identification" with living things, human and inhuman. "If the bush soul is a tree, the tree is presumed to have something like parental authority over the individual concerned… An injury to the bush soul is considered an injury to the man" (p. 7), and vice versa. On this and related themes of eco-psychology and eco-phenomenology see David Abram, *The Spell of the Sensuous: Perception and Language in the More Than Human World*, Pantheon Books, 1996; and *Becoming Animal: An Earthly Cosmology*, Vintage, 2011; and Erazim Kohak, *The Green Halo: A Bird's Eye View of Ecological Ethics*, Open Court, 2000.

8 The Lakota phrase *Mitákuye Oyás'iŋ* describes reality by addressing it as "All My Relations." All humans, all animals, all plants, all the waters, the soil, the stones, the mountains, the grasslands, the winds, the clouds and storms, the sun and moon, stars and planets are our relations and are relations to one another. Brian McKlaren describes a similar vision of multilateral symbiotic interdependency (rather than vertical hierarchy): "We aren't ruling from the heights of a great top-down pyramid (as) generals under King God in the divine chain of command. We aren't given by our rank a carte blanche to dominate, oppress, exploit, and exterminate everything below us. No, we aren't at the top of anything; we're simply at the tip of one small branch of a very huge, verdant tree, and all created things are our grandparents, cousins and siblings" (*The Gallapagos Islands: A Spiritual Journey*, Fortress Press, 2019, p. 214).

9 Lisa and Wolf Wahpepah in conversation with Michael Kearney, "Listening to the Earth" in this volume, originally recorded on guestbookproject.org. Wolf Wahpepah adds this story from his native Iroquois tradition: "The Great Peacemaker was shown a vision of the Tree of Peace where he saw that the diversity of the people could still be unified. And even though there were sub-groups and many tribes, the Creator showed him that they could form a union like a tree growing with different branches. And these different branches would be the different nations of people that lived, all joined by one trunk, and that if they were in proper relationship to one another in a system that wasn't vulnerable to corruption, that it would be a system based on fairness. By incorporating and adopting more and more tribes, his intention was to literally end all warfare on the North American Continent, on Turtle Island." See also Randy Woodley on the sustained ecological hope of indigenous First Nation communities in North America. "Their real hope recognizes that Earth endures and that we can still do enough to reverse the damage done. After all, the Earth is much stronger and more resilient than any human being. Although human beings are a part of the Earth, we may be the most expendable. This gives me pause—as well as a much longer view of our history and our future. I think Mother Earth is going to be alright in the end. I just hope we will be here long enough to see it. Although it might make us feel pretty insignificant, another way to turn the phrase is this: 'We are still here… for now. But the Earth remains forever'" (*Becoming Rooted: One Hundred Days of Reconnecting with Sacred Earth*, Broadleaf Books, 2022, pp. 57–58).

10 Robin Kimmerer, *Braiding Sweetgrass: Indigenous Wisdom, Scientific Knowledge and the Teachings of Plants*, Milkweed Editions, 2013, pp. 128–140.

11 The earth is a process of dying as well as living, of waste, pain and putrefaction as well as wellness and growth. It comprises a complex holistic ecosystem harboring predators as well as nurturers, viruses as well as vitamins, parasites as well as hosts. The cycle of life-death (*eros-thanatos*) is integral to nature, as it is to our unconscious psyches, and we can only fully care for the earth when we integrate the shadow as well as the light. Symbiosis holds both. Hence the need to avoid sentimentalizing nature—casting it in idealized images—or demonizing it with fearful fantasies. Nature remains,

deep down, an unknowable enigma, at best a numinous mystery. See Brian Treanor on the deep perplexity of many nature writers—e.g., Henry David Thoreau, Nan Shepherd, Annie Dillard—before these complex paradoxes of natural life ("Preserving the Wilderness Idea," forthcoming in the *Hedgehog Review*, 2024; and "Thinking Like a Jaguar: Carnal Hermeneutics, Touch and the Limits of Language" in *Anacarnation: Returning to the Body with Richard Kearney*, ed. Brian Treanor and James Taylor, Routledge, 2023, pp. 1–11).

12 I am grateful to astrophysicist Leon Golub and Anne Davenport for these examples. On the discussion of symbiogenesis in the natural life sciences see note 3 above. See also the work of Lubna Dada, an atmospheric scientist who studies how trees and plants contribute to cloud formation, cooling "cloud seeds" when they are stressed by heat.

13 Another philosophy informing our "hosting earth" vision is Celtic panentheism which sees divinity as a river (*deus currens*) running through all living things. See our discussion of Eriugena, Duns Scotus, and Pelagius in "My Way to Theopoetics through Eriugena" in *Literature and Theology*, Vol. 33, No. 3, 2019, pp. 233–240. Celtic theologies of nature chime with later panentheist spiritualities ranging from Saint Francis's *Canticle of Creation* to the vitalist nature mysticisms of Bruno of Nola and Hildegard of Bingen. The Celtic thinker, Pelagius, for example, celebrated God's presence in nature thus: "Look at the animals roaming the forest/Look at the birds flying across the sky/Look at the insects crawling in the grass/Look at the fish in the rivers and sea/Look at the great trees of the forest/Look at your crops and plants/God dwells in them all." And Pelagius adds: "There is no creature on earth in whom God is absent… When God pronounced that his creation was good, it was not only that his hand had fashioned every creature; it was that his breath had brought every creature to life… The presence of God's spirit in all living things is what makes them beautiful; and if we look with God's eyes, nothing on earth is ugly" (cited by Richard Rohr Center for Action and Contemplation, 2018). On the formative influence of Eriugena's Celtic philosophy on subsequent Irish thought, see "Borges and the World of Fiction," an Interview with Jorge Luis Borges by Seamus Heaney and Richard Kearney in *The Crane Bag*, Vol. 6, No. 2, 1982. Another important Celtic thinker, Duns Scotus, acknowledged the "mystery of incarnation" in the particular thisness (*haecceitas*) of each created thing. Such a Celtic-Franciscan vision sees "the visible world as an active doorway to the invisible world… Matter is, and has always been, the hiding place for Spirit, forever offering itself to be discovered anew. Francis and his companion St. Clare carried this mystery to its full conclusions… (knowing) that the beyond was not really beyond, but in the depths of *here*" (Richard Rohr, Centre for Contemplation and Action, Oct. 3, 2023). This Scotist-Franciscan view of nature would later find powerful expression in the poetics of Gerald Manley Hopkins who spoke of a sacramental inscape in all living things (see his classic nature poems, "Pied Beauty" and "When Kingfishers catch Fire"). See also the Franciscan panenthesim informing Pope Francis' encyclical on nature and climate, *Laudato si'*, outlined in Sean McDonagh, *To Care for the Earth: A Call to a New Theology*, Chapman, 1986. Other contributors to our volume echo this panentheist perspective in their exploration of panpsychic philosophy, mystical theopoetics and ecological mindfulness—in particular, Fanny Howe, Christopher Yates, Sean McGrath, Chandler Rogers, and Melissa Fitzpatrick as well as environmental artist-thinkers Jane Marsching, Sheila Gallagher, and Edward Casey.

14 See Michael Kearney, *Becoming Forest*, pp. 35–36, 131–132, 201, where the author recommends "mindfulness breathing" as a basic practice for reconnecting with our natural environment in a spirit of mutual nurturing and "deep resilience for uncertain times." In similar vein, see Leah Schade on the symbiotic dialectics of inhalation-exhalation: "As you breathe in, you are taking in oxygen, which is released by trees and all green-growing things. As you breathe out, you exhale carbon dioxide, which in turn is being taken up by trees… You are as solid as the earth and made from the same atoms of carbon, oxygen, hydrogen, and nitrogen that make up the earth… All the elements that

make up your body came from stars that exploded millions of years ago... connecting us to the air and to plants, to the earth, to waters and the sea, to the animals, and to the stars... We exist within an interconnected web of relationships—brother-sister beings with the rest of life" ("Kinship with Creation," in *Rooted and Rising: Voices of Courage in a Time of Climate Crisis*, ed. Leah D. Schade and Margaret Bullitt-Jonas, Lanham, MD: Rowman & Littlefield, 2019, pp. 76–77. See also Jean-Louis Chrétien who, like the French mystic Claudel before him, takes the paradigm of "cosmic respiration" as central for all flourishing interpersonal relations, connecting self and other. By breathing in and out, he writes, "I am possible only through another, from another, and by taking that other into myself. The model of autarchic self-reliance and of self-growth whereby I would aim at thriving by myself is really a model of asphyxia and death" (*Spacious Joy*, Trans. Anne Davenport, Rowman and Littlefield, 2019, p. 170).

Such a phenomenology of symbiotic respiration points beyond the individualism of the Anthropocene.

15 See Gaston Bachelard, writing of the ecological "milieu" between human and arboreal nature: "It breathes me. The world comes to breathe within me; I participate in the good breathing of the world" (*The Poetics of Reverie*, Beacon Press, 1969, p. 179).

16 See the Intergovernmental Science-Policy Platform on Biodiversity and Ecosystems Services, 2019, cited by Michael Cronin, *Irish and Ecology*, Foras Na Gaeilge, 2019, p. 36.

17 See Timothy Morton, *Being Ecological* (Pelican, 2018), discussed by Michael Cronin, *Irish and Ecology*, pp. 47, 51–56. Morton believes that humans have been traumatized by being sundered from non-human beings and calls for new modes of reconnection to places and practices of nature. Cronin sees such practices of reconnection as deeply tied to a recovery of lost or threatened linguistic and ecological cultures, which need to be retrieved if we are to respond to our anthropogenic climate crisis and so attempt to give a future to our past (pp. 51–56). Such insights are in keeping with a whole generation of contemporary nature writers ranging from John Moriarty (*Nostos*), Mary Oliver (*A Thousand Mornings*) and Wendell Berry (*The Gift of Good Land*) to Manchán Magan (*Thirty Three Words for a Field*), David Wood (*Reoccupy the Earth*) and Robert MacFarlane (*Landmarks*), all of whom explore the intimate symbiotic relationship between landscape (nature as place) and language (poetics).

18 See Guestbookproject.org and our discussion of this theme in previous publications such as *Phenomenology of the Stranger* (with Kascha Semonovitch), Fordham UP, 2011; *Hosting the Stranger* (with James Taylor), Continuum, 2009; and *Radical Hospitality* (with Melissa Fitzpatrick), Fordham UP, 2022.

19 What the ecological emergency brings home to us is that nature can no longer be considered something external to us. Instead of priding ourselves as a species apart, dominating other species on this planet, we should humbly acknowledge that we are one species amongst others living in a web of mutual interdependency. This realization of our symbiotic relation to nature calls in turn for a rapprochement between the human and natural sciences which have functioned as rivals for centuries, the former devoted to matters of psyche, spirit, and the arts while the latter gravitated toward a "naturalist" description of empirical experience as a composite of measurable and computable data. The climate crisis reminds us that human nature and non-human nature can no longer be dualistically segregated in this manner, even for the purposes of knowledge. And this calls for a new epistemological hospitality between the sundered human and natural sciences and, by extension, a radically new politics of sovereignty: "a power-sharing where we look at non-human agents (water/temperature/soil) as constituent parts of a common world... a sort of post-human, post-nationalist sovereignty that sees territorial integrity not based on separation and exclusion but on interdependency and inclusion" (Michael Cronin, *Irish and Ecology*, p. 69. See also our essays on postmodern sovereignty in *Post Nationalist Ireland*, Routledge, 1998).

20 Edge walkers, according to Victoria Loorz, are those called to the thresholds—the edges between the polarized spaces which most people inhabit during the reign of the Anthropocene. The edges between biosystems are called *ecotones*, marking thresholds which contain the most biodiversity and therefore are the most resilient. Loorz calls for a time "when the edges we inhabit will start to redefine the center. And we will need to lean on and learn from one another as we, together, engage in the work of that redefining. Each of us is characterized by our own unique gifts, communities of influence, and a particular bio-region. But we cannot behave as silos. The more diverse our relationships are, the more resiliently we can hold our own individual edges" (p. 15). Loorz applies this edge walking to interspiritual symbiosis as a comparative theology of the earth. Citing a gathering of spiritual leaders from different traditions, she notes that what they shared most in common was *life on this planet*. She observes a deep link between interfaith compassion and eco-hospitality: "We talked about how our faith traditions could connect us with the actual soil and water and creatures of Earth. And how that connection could be a spiritual foundation for the environmental movement. What I remember most was a golden thread of mystical connection with divine presence that all of us expressed in our relationships with the natural world. Even in our diversity, we all felt that we had more in common with one another—edge walkers from other traditions—than we did with people more firmly planted in the center of our own faiths" (p. 14). She concludes with a plea for a new generation of edge walkers who would refuse to follow the dualist furrows of the Anthropocene, acknowledging instead that "at that edge, spirituality and nature are in unbroken relationship" (Victoria Loorz, *Church of the Wild: How Nature Invites Us Into the Sacred*, Broadleaf Books, 2021, pp. 14–16). On this idea of edge walking between spirituality and nature, see also Benjamin Webb, "In Search of Our Fugitive Faith: Terry Tempest Williams," in *Fugitive Faith: Conversations on Spiritual, Environmental, and Community Renewal*, Orbis Books, 1998. On the panentheist co-belonging of earth and divinity see the discussion by Franciscan sister and scientist, Ilia Delio, on nature as a symbiotic holarchy rather than a top-down hierarchy (*The Unbearable Wholeness of Being: God, Evolution and the Power of Love*, Orbis Books, 2013, pp. 128–129, 130, 131). See finally the work of Maxine Greene and Jennifer Ayres on sacred geopoetics and ecotheology in *Inhabitance: Ecological Religious Education*, Baylor University Press, 2019; and Karen Armstrong, Sacred Nature, Anchor Books, 2023.

21 Maurice Merleau-Ponty, "Eye and Mind" in *The Primacy of Perception*, Northwestern UP, 1964, p. 167. Merleau-Ponty cites the painter Klee who attested to the reversible relationship between the seer and the seen: "In a forest I have felt many times over that it was not I who looked at the forest. Some days I felt that the trees were looking at me, were speaking to me… I was there, listening… It becomes impossible to distinguish between what paints and what is painted" (p. 167). Merleau-Ponty bases his phenomenology of reversible double sensation on the experience of each embodied subject immersed in the "flesh of the world." We are bodies inhabiting the greater body of the earth in a two-way relation. We are in nature and nature is in us. Or as Paul Valéry wrote, language is above all else "the very voice of the trees, the waves and the forest." The earthly biosphere we humans inhabit is, deep down, an interworld (*entre deux*) of mutual intertwining and entanglement. A "chiasm" wherein what I touch is at the same time what touches me, just as what I see is what sees me. Gaston Bachelard also writes of this double tactile vision: "Between the seer and the world there is an exchange of looks, as in the double look between lovers… Everything I look at looks at me" (*Poetics of Reverie*, p. 185). I have endeavored to develop this phenomenology of double sensation in a number of recent works including *Touch: Recovering our Most Vital Sense*, Columbia UP, 2021, pp. 45–52, *Carnal Hermeneutics* (edited with Brian Treanor), Fordham UP, 2015, Chapter 2, and "Anacarnation: Recovering Embodied Life" in *Anacarnation: Returning to the Body with Richard Kearney*, ed. Brian Treanor and James Taylor, Routledge, 2022.

Part I

On Climate Justice: An Interview with Mary Robinson by Richard Kearney

Richard Kearney: Welcome Mary. You have been engaged at the highest inter-national level with the recent debates on our climate emer-gency. Your book, *Climate Justice*, is yet another step in your campaign to address this most important moral, political, and existential issue of our time. Namely, the ecological question concerning the survival of not just our human species but all living species and the very planet earth which hosts us. Your book is essentially a book of narratives from climate activists throughout the world. So my first question to you is this, why did you choose to privilege the storytelling method?

Mary Robinson: First of all, I realized that the world was not able to understand our current climate crisis because of the way most people were hearing about it. So I decided that we had to move to story-telling to make the message more concrete and available. In my work, I wanted to include stories from the real "experts" who know the most about the impacts of climate change—not just the scientists and statisticians (though these are essential too) but ordinary people who are desperately struggling to defend their living communities from the absolute reality of climate threat and extinction. I was determined to be at once as global and as particular as possible. Hence, I was keen to include stories from the United States, not just in the Southern Hemisphere which has been experiencing the appalling impli-cations of the climate crisis for decades but also the Northern Hemisphere which has often considered itself more safe and removed from the immediate danger—an illusion of course.

When I was writing the book, it was particularly difficult to get stories from North America. In Copenhagen, I attended my first climate conference where I met a girl from Mississippi, Sharon. Sharon lived on the wrong side of the tracks where she had her lovely salon and her home. Living there, Sharon learned what it is to suffer real poverty. In Copenhagen, we sat

DOI: 10.4324/9781003456940-1

together and she talked about the humility of having her entire life wiped out by Hurricane Katrina. She was an accidental activist who was fighting back to change her circumstances and those of others around her.

The other story I chose from the United States was in Alaska—that of Patricia Cochran. I first met Patricia at a big event in London for women that Tina Brown had organized. It was one of Tina's great events and Patricia was there as a climate scientist from Alaska. I spoke to her and was so impressed by the way she told her story as a scientist. She is an indigenous scientist who felt this enormous connection with the earth and was watching it crumble and get destroyed. She had a great social conscience about it all and so these were two of the stories I discovered at the time which I wanted to relate in order to change awareness in the developed as well as the developing world. If I was writing now, I would not have had trouble finding stories of the more advanced world but I was deliberately choosing stories about injustices committed by the developed world.

I also chose most of my stories from women: 80 percent of the food in the developing world is produced by women yet they own far less of the land. Women get much less access to financial credit so I wanted to bring that out and subsequently embrace another way of telling stories. When I moved to a podcast called "Mothers of Invention," I initially did it with a young Irish comedian based in New York, Maeve Higgins. She was eight years old when I was elected President but is now well known, especially since she made films. When we met in New York, we hit it off because she was half respectful of me and there was half humor between us. I remember one of her first questions on air to me was, "when you went out on your first date, who paid?" Even though Maeve did not know anything about climate, she knew how to ask questions. We kept interviewing special people together and many people had their questions asked in a way that helped them to understand the climate issue.

But, the reason I mention Maeve is that we had a byline in our podcast that climate change is a man-made problem that requires a feminist solution. I always try to explain how climate change is a manmade problem: (i) it is generic, (ii) we are all responsible, and (iii) a feminist solution is increasingly one that we should be embracing. The feminist solution is problem-solving; it is inclusive. It is listening to all the voices, it is multifaceted and has grown significantly out of the feminist movement. But I am happy to say that now more and more men acknowledge it. They are bringing themselves together around a feminist approach.

RK: At this very tardive stage in the game is climate justice still doable? Or is it just a utopian aspiration?

MR: I think the window is closing, four years after I have written the book. That's the worrying thing and I'm so conscious of it almost every morning when I wake. I am watching things carefully. On one of my television segments, there is a climate action tracker which is a very good NGO that records climate movement and the pledges that governments make to businesses. They talked about what the private sector was doing, and what investments were being made, and the climate action tracker added up all the pledges to calculate the total impact. In Paris at COP26, I was the envoy of the UN Secretary General at the time. There, I was witness to the planning by the French presidency which was quite intensive. President Obama and President Xi needed to decide that there would be a climate agreement when they met about it in China. Indeed, it was telling that the most significant measures were taken because of pressure from the indigenous peoples, civil society, and young people who marched in the streets of Paris. The mantra of the march was "1.5 to stay alive! 1.5 to stay alive!" Towards the conclusion of the conference, a high-ambition coalition was formed and the head of that coalition was Tony de Brum, the Foreign Minister of the Marshall Islands. Then the EU joined that coalition, then the US, followed by other countries. The sole purpose of the high-ambition coalition is to ensure that the goal adopted by the Paris Agreement stays below 1.5 degrees in increase.

The global climate scientists were asked by the Paris Agreement study to bring in a special report on the difference between a 1.5 and 2 degrees increase. If we were to hit an increase of 1.5 degrees, all the coral reefs will probably disappear entirely, the Arctic ice might well disappear and the permafrost would melt. These factors would all accentuate the climate tipping point problems, so the scientists at the time said the whole world has to stay at 1.5 degrees increase. To do that, we need significant political will. The Report states that we have 12 years to reverse the damage, and now—Summer 2022—we have less than eight.

I remember talking a few years ago to the International Energy Agency (IEA), a very conservative body that represents oil and gas and all the other agencies, and it had just taken on a new Turkish director. He invited me to come and talk to him about climate justice but I said to myself this organization is part of the problem. It has now completely changed and if you read the important reports of the international energy agency, they are encouraging people to get out of fossil fuels as quickly as possible. It is just amazing how significant the power of the fossil fuel lobby is. It is incredible to see how awareness can alter minds and politics.

RK: We can't solve the climate crisis without addressing the issue of poverty so could you maybe say something about economics and the practicality of actually doing something, given the pressures? Surely economics and ecology are intimately connected when it comes to addressing the climate emergency?

MR: I think it's important that we grasp what I often call the interrelated layers of injustice. The first is realizing the problem. As somebody who spent seven years from 1990 to 1997 as President of Ireland, I spoke about the environment but I didn't mention climate. The climate wasn't visible then and nobody was speaking about climate change. Then I went to the UN and there I learned that climate is a very serious issue. I got to discover another world of human rights, gender equality, and the rights of people with disabilities. I realized I had missed this really important connection which then became my passion.

Over time, I discovered that climate change was impacting far more of the African countries. It was directly impacting indigenous peoples and the poorest parts of richer countries as well. People living in richer emerging economies were suffering and they were the least responsible because they were not driving cars, not using central heating, not doing major manufacturing and they were the black and brown and indigenous peoples of our world; so it's a kind of racial injustice as well. Within this realm, gender injustice also became striking as women do have different social roles, different powers, and sometimes different land rights and different roles in society. All are subject to all of these inequalities. And yet they have to also put food on the table, go further to fetch water in droughts etc. The third layer is intergenerational injustice. There are millions of children and young people in the world who haven't got the political power to make decisions yet and those who have the power are not making the right decisions. But the young people are very smart and they're very connected now and their pressure is helping. They're becoming very angry and frustrated.

The problem for developing countries who pledged in the Paris agreement is that they wanted to go green as soon as possible. But, in order to go green you need upfront expenses, you need serious investment and they haven't got it. It's very hard to get heavy investment for developing countries. Indian

Prime Minister Modi managed to get quite a bit of funding out of the United States and India has gone significantly solar. Climate change has essentially happened throughout my lifetime. It happened in the 1950s when we were stupid enough with our industrialized age not to realize that we were taking ourselves out of the Cinderella period of the world. Two thousand years ago, the Colosseum was never too hot, never too cold and civilization flourished. We humans, over a short period of time, have affected what has happened to the earth.

Moreover, what worries climate scientists more than anything else is that nature is not predictable. At the moment, nature is our great friend, the oceans sink huge amounts of carbon, and the rainforests give us abundant resources. But we're killing everything due to our lack of awareness. Perhaps the planet will survive without us and go back to the Holocene. But we will be gone from the Earth because we will have made it unlivable for ourselves. It is really hard to believe how difficult it is to get politicians and their policies to keep the focus on climate change and this also shows the power of the fossil fuel lobby.

RK: This reminds me of something that came up at our "Hosting Earth" conference when Jane Marshing, an environmental artist here in Boston, warned about becoming obsessed with a climate apocalypse. We need to get the hard facts out there, of course, and know how horrible our current situation is, but then we've got to start thinking otherwise. We can't give in to apocalyptic thinking lest it become a self-fulfilling prophecy. The poet Fanny Howe also spoke here yesterday about the paradox of the "wonder horror." We need to register the horror but never lose sight of the wonder of nature.

MR: I agree. I tend to encourage people—when I'm talking about Climate Justice—to take three steps. The first step is to make the climate crisis personal in your own life. You make it personal insofar as you're doing something new today and tomorrow that you weren't doing yesterday. Perhaps it's recycling more carefully or changing your diet or whatever small thing it might be. But what is important is that you own your responsibility.

Secondly, get angry with those who aren't doing enough. That's government, business, investment, the cities. Try and see who's doing what. Can you help by joining the many great programs and NGOs around the world that are putting pressure on those in power to make a difference? And the third thing—that's the most important—is to imagine the world we are hurrying towards. Unless we're doing that, we're not going to have the collective sense that we must do something. People change their minds through their hearts. We have to get a real feeling from the heart that we must do something so that tomorrow will be better—and certainly a lot healthier. And lastly, energy is going to be free, because we will be relying upon the natural elements of sun and wind and we're already seeing that it's a lot cheaper, at least in the long run.

RK: Another participant at today's conference, the environmental thinker David Storey raises the question of climate change and literature. He cites the case of how Kim Stanley Robinson's novel, *The Ministry for the Future*, became such a sensation that he was invited to the COP talks to address several groups. In the book, the author considers what a technologically, politically, and economically realistic solution to the problem would look like. Such that by 2050 or so, humanity more or less gets its arms around the climate problem. But the irony is that Kim Stanley Robinson was most pursued by central bankers and big-shop finance people. Because—spoiler alert—around 2050, it's the central bankers who save the world because they re-engineer the global financial system to incentivize decarbonization of the global energy economy. It's fascinating to me because you think about speculation in financial markets and speculation is what science writers do. And central bankers and financiers were turning to a literary imagination because they didn't know how to do it themselves.

MR: The sooner the storyteller gets the central banks together the better!

RK: The question of nuclear energy has also come up a lot in our 'Hosting Earth' conference discussions. This is, as you know, a very controversial issue in environmental circles. Made worse in the current context of the war in Ukraine where Europe and the West are cutting their dependency on Russian oil and gas. Germany, for instance, relies greatly on natural gas since they denuclearized after the Chernobyl and Fukushima disasters. Because nuclear is the only reliable baseload form of power that doesn't face the intermittency problems of renewables, what role can it play in the energy mix of choices that countries must make going forward?

MR: I'm not a fan of nuclear and I know that building nuclear plants has been extremely expensive; and then of course there is the waste disposal problem. Frankly, because we're in such a short window of time, I know that it doesn't add to the problem with emissions and therefore I'm hearing about how some small nuclear technology could be viable. I do feel, however, that the real focus should be on renewables and that's what climate scientists are saying to us as well. Because they're getting so much cheaper and the battery retention so much more effective and because they're linking with the grids much more effectively, we should go with those technologies that have us move with mother earth in a much smoother way, rather than use a technology that has its downsides. Though I have somewhat softened on the issue because we have not yet determined how to take the carbon emissions out if we keep putting them in at our current rate. I really fear when we hear what the scientists say about the difference between 1.5 degrees and 2 degrees—and we're already above one degree—it's very urgent and we need a different economic system.

RK: Are you an optimist or a pessimist at this stage? How do you weigh the rival claims of hope and despair as the science becomes more and more alarming?

MR: I am conscious that I am saying some very tough things and I'm worried that I'm not fulfilling Mandela's mandate. Can I tell my Tutu story? I was on a panel with Archbishop Tutu some 15 years ago in New York. It was a "social good" panel with young people on their phones and whatnot. I was an Elder at the time and he was Chair and neither of us was terribly "social-media conscious"; but when Archbishop Tutu is in front of young people—he loves them, he throws up his arms with enthusiasm. We were being moderated by an American journalist and she said quite sharply: "Archbishop Tutu, why are you such an optimist?" He turned to her and said, "oh no, I'm not an optimist, I'm a prisoner of hope." And you know those words struck home to me because I could see what he meant and we talked about it afterward. You have to be a prisoner of hope because that gives you the energy to take the next step. If you talk about climate in an apocalyptic way, then all the energy goes out of the room and people actually put their heads down and feel hopeless and just go on with their lives regardless. Whereas this is saying that we all are destined to want to do something. We move with the heart rather than with statistics and the head. And we get it done. That is being a prisoner of hope.

RK: So we end with the old question, what is to be done? In spite of the frustration and anxiety many of us here at this conference feel, you have given us three things to do. First, to make the climate crisis more personal. Second, to get angry with those in positions of power and influence who are not doing what they should be doing; and third, to try to identify and support NGOs and other organizations that are creating grassroots green teams which are working for change.

MR: It is true that there are a lot of NGOs working on conservation, on green nature, on clean energy, on women's impact on climate change (e.g. Mothers Out Front), on helping vulnerable developing countries in different ways. So it depends on where you want to bring change—there are many options and opportunities for young people to get involved and have a real impact.

RK: This conference is taking place at a place of learning and I would like to ask, finally, what advice you might have for universities throughout the world on the vexed question of divestment?

MR: Yes. Universities are leaders and they should be leading from the front in every sense on this critical issue. I have for a long time supported the idea of universities with endowments committing to divestment. I think my own alma mater Harvard is finally making some progress now. I also approached Trinity College in Dublin to start a process of divesting a few years ago when I was Chancellor there at the time. It was the students who started a real protest and the president was caught in the middle and decided to put himself on to the Wallace Foundation. The truth is that universities find it difficult and there is a lot of blended finance so it is not easy work. But what you do is make the commitment and then work out a way to live up to it. All of the universities should be doing that now. No endowments should be invested in what is destroying this planet. I do accept that educational authorities have their employees and pensions to consider, and other complicated things. But many universities are endeavoring to find ways and are not regretting it. And some are making moves to invest in renewables. That is an important part of the story because we need as much investment in renewables as possible. I am conscious that I have been saying some tough things and am worried that I may not be sufficiently fulfilling Mandela's mandate.

RK: Well Mary I think you represent a courageous and timely mix of Tutu's call for hope against hope and the moral call to talk tough when it comes to responding to the cry of our suffering earth and fellow species. As "prisoners of hope" we are all deeply grateful to you.

Boston College Earth Day Conference April 2022

Part II

Poetics of the Earth

Chapter 1

Poetics of Earth: A Colloquy

Melissa Fitzpatrick, Brian Treanor,
Catherine Keller, and Jason Wirth

POETICS OF THE EARTH
Melissa Fitzpatrick

We have a matter of years to reverse our current course, which is terrifying considering how much so many love the luxuries and conveniences of that course—and how suffused we are in a system that sometimes gives us no choice but to want that convenient course. How do we change our priorities and values? And persevere in the face of so much alarming, overwhelming, and complex news? What and how should we create in hopes of a new economic order? The "us versus nature" paradox modernity has left the West with prompts us to probe how we can overcome an understanding of the earth and its inhabitants as the kinds of others who are below us, to be used and abused, and thus shift toward an understanding of the earth as full of other-than-human persons. That is, an understanding of the earth as home to a wealth of diverse personalities who in fact constitute our very existence.

With *poesis* as *creating* and *making* in mind, it seems vital to think through the idea of hospitality as a conversion of sorts, in the sense of being marked by a change of heart. It involves a turnaround of the mind in a way that makes it more generous and welcoming—more attuned to the various others "outside of" oneself. In Robin Kimmerer's brilliant book, *Braiding Sweetgrass*, she starts the work of sketching what this change of heart might look like. In the chapter titled "The gift of strawberries," she reminisces about the field of wild strawberries that was behind her home in upstate NY—and the gifts of pie they would make for her father. Quoting Kimmerer:

> This is the fundamental nature of gifts: they move, and their value increases with their passage. The fields made a gift of berries to us and we made a gift of them to our father. The more something is shared, the greater the value becomes. This is hard to grasp for societies steeped in notions of private property, where others are, by definition, excluded from sharing. Practices such as posting land against trespass, for example, are expected and accepted in a property economy but are unacceptable in an economy where land is seen as a gift to all.

DOI: 10.4324/9781003456940-3

From the viewpoint of private property economy, the "gift" is deemed to be free because we obtain it free of charge, at no cost. But in the gift economy, gifts are not free. The essence of gift is that it creates a set of relationships. The currency of gift economy is, at its root, reciprocity. In Western thinking, private land is understood to be a "bundle of rights," whereas in gift economy property has a "bundle of responsibilities" attached.

(*Braiding Sweetgrass*, pp. 27–28)

Kimmerer I think rightly suggests that "A great longing is upon us to live again in a world made of gifts" (*Braiding Sweetgrass*, p. 32). Many people have (successfully) lived and in fact do currently live this way—that is, with an understanding of the world as being made of gifts. But for those who do not, how can they or we begin to make that creative turn? How do we make and create in light of the great longing for a world made of gift? And what mode of being lies on the other side of that turnaround of the mind? On the other side of conversion? Who is this new generation (Generation Alpha)? Since debt—perhaps too linked to transaction and the private property economy—is not helpful here, how should we understand the type of binding relation that is bundled into the gift in *praxis*?

It seems clear that the generation of tomorrow will be one that is willing to embody a new economic order—that is, an economic order that recovers the essentials dimensions of who we most fundamentally are, while being daring enough to create something completely new, attuned to the complexities of the present.

HOSPITALITY AND THE EARTH: HOLDING BACK FOR THE OTHER
Brian Treanor

Introduction

In early autumn 2003, Timothy Treadwell was camping in Katmai National Park at the end of his thirteenth summer observing, living among, and increasingly interacting with the grizzly bears (coastal brown bears) that make their home there.

He had overcome a troubled history of substance abuse in part through his interest in and enthusiasm for grizzly bears; and it would not be an exaggeration to say that the bears, and his complicated relationship with them, saved Treadwell's life. Over the course of his many visits to the Katmai Treadwell became increasingly cavalier, and eventually careless, about the inhuman otherness and power of these apex predators. He thoroughly anthropomorphized the bears, ascribing to them human personalities, motives, and goals. He named them and spoke to them in the high-pitched, sing-song manner in which adults over-emote to infants or domesticated pets. He sometimes spoke of himself as a bear—intoning, "I am a grizzly"—and began to approach them more closely, even touching them. He writes of defusing an encounter with one bear by telling it, "I love you," and of kissing another bear, "Peanut," on the nose after it had licked his hand.[1]

Treadwell spoke and wrote sincerely about his love for the bears. He spoke of being more at home in the Katmai than he was back in Los Angeles, referring to his seasonal journeys north as a "homecoming." He also believed, against evidence, that the bears were endangered, and that the only thing protecting them from poachers and other harm was his presence. His diatribes against the Park Service, wildlife biologists, and others who criticized his behavior seemed both conspiratorial and, occasionally, unhinged. Out of his concern for and identification with the bears, Treadwell swore off carrying bear spray and ringing his camp with a basic electric fence—both common, and recommended, practices. He seemed to believe that his love and respect for the animals would be adequate protection, although he repeatedly acknowledged the reality that they could, and might, kill him.

Unfortunately, Treadwell's final visit ended tragically when he and his companion, Amie Huguenard, were attacked at their camp the evening before they were to fly out for the season. Although the details of the event remain unclear, a six-minute audio tape found amid the wreckage of the camp seems to suggest that Treadwell exited the tent and confronted what was probably a large male grizzly. The bear attacked Treadwell, who—in keeping with one established strategy—tried to "play dead." The bear left, only to return shortly thereafter. Treadwell was killed and consumed; Huguenard was killed, partially consumed, and stashed for retrieval.

Treadwell claimed to feel at home in the Katmai and felt that the resident bears accepted and welcomed him. He sought to reciprocate, relating to the bears in a way that seemed to him loving and respectful. But whatever one thinks of his story, Treadwell's ultimate fate makes it clear that he profoundly misread the reality of the situation. And while in one sense the bears of the Katmai did save Timothy Treadwell's life, they also ended it savagely and absolutely.

Hospitality as Home-Sharing

What would it mean to be hospitable to the Earth? Or to receive hospitality from it?

Needless to say, a great deal hinges on just what we mean by "hospitality." Since the early Platonic dialogues, philosophers have recognized the value of defining terms. We will not get very far in thinking about justice (the *Republic*), piety (*Euthyphro*), friendship (*Lysis*), or courage (*Laches*) if while discussing them we mean radically different things by justice, piety, friendship, or courage. Or, in a modern idiom, we will not get very far in thinking about whether or not particular systems are racist if we are operating without a shared sense of what "racism" means. Absent that shared definition, it is entirely possible for a single thing to be racist and not-racist at the same time. Unless we agree about the subject we are discussing, in at least a provisional way, we will have a hard time explaining ourselves or sorting out whether and where we disagree.

I think we run into a similar problem when we think about hospitality and the Earth. Once we get away from the thoroughly commodified "hospitality" industry, the first things that come to mind when we think of hospitality are domestic scenes: lived-in, well-worn, multi-generational homes; warm hearths, aromatic kitchens, and holiday meals; amiable interactions with neighbors; and so on. A second image, building on the first, has to do with the ways in which hospitality is offered not to friends and neighbors, but rather to the "widow, orphan, and stranger." Here we might think of our own experiences of being welcomed with hospitality, or of offering hospitality to others. I've been taken in out of the rain by drivers while hitchhiking across Ireland, spontaneously "invited"—abducted, though to my delight—to watch football matches in the foothills of the Atlas Mountains, and treated to tea by monks while living in Japan. In each instance, my hosts saw someone who was obviously out of place, and opened their car, table, or home in an act of hospitality.[2]

In light of these examples, which seem somehow paradigmatic of hospitality, I have argued elsewhere that hospitality is fundamentally a virtue of the home, and can only really be offered by someone who is implaced to someone who is displaced.

> Only an implaced person can be hospitable. A displaced person, qua displaced person, can be generous, can be the giver of gifts, can be forgiving, and can be responsible, but she cannot be hospitable because she cannot give place to an other... When the host ceases to be a host (as when she herself is displaced) or when the guest ceases to be a guest (as when she becomes a naturalized citizen or member of the family), we can no longer speak of hospitality.[3]

Thus, for example, it would be very odd to say that I can be hospitable to my wife—that is, that I can graciously open my home to her—precisely because she is my wife, and my home is by definition her home. Likewise, while the United States can be, and at its best has been, hospitable to the tired, poor, wretched, and homeless referred to in Emma Lazarus's "The New Colossus," once those people have become naturalized citizens, hospitality no longer applies, because they and their descendants belong here as much as the descendants of the Mayflower.[4] When a person marries into my family, he or she becomes family. If an immigrant comes to my country and becomes a naturalized citizen, he or she is an American. To suggest that I can continue to be hospitable in either case implies that the person—my son-in-law, the naturalized citizen in my neighborhood—is somehow out-of-place, *that they do not belong*. There are many virtues we can and should exercise in our relationships with such people; but hospitality is not one of them.[5]

Call this the "narrow" sense of hospitality: the act of opening one's home or one's place to the stranger, and helping her to feel "as if" at home herself.

If we think of hospitality in these terms, however, talk of "hosting Earth" must be viewed with significant skepticism. On the one hand, if we talk about the Earth hosting humanity, we risk implying one of two things, either of which is disastrous. The first is the implication that humans are guests here. Our *real* home is elsewhere; we are only visitors on this blue marble. This is a problem because such a view encourages, or at least facilitates, mistreatment of the non-human world. It is no surprise that people treat rental cars and Airbnb flats poorly. A second difficulty with the Earth hosting humanity would be the implication that humanity is a parasite of sorts, essentially rather than accidentally destructive to the host.[6] Here again, thinking of the Earth as host to humanity seems to suggest that our very presence is in some way unnatural or noxious. When we speak of the Earth hosting humanity, we court either anti-humanism (humans don't belong because they are parasites the host should eliminate) or Gnostic dualism (humans don't belong because their real home is elsewhere).

On the other hand, if we speak in terms of humanity hosting the Earth, we run into the problems raised above. How can I play host to that which is my home? Can I be hospitable to my house? A young child cannot play host to her mother, first because the mother is also at home in the daughter's house, and second because the implacement of the child is dependent, temporally and materially, on the implacement of the parent. Likewise, I cannot host the Earth, which is both my home and the condition for the possibility of my own implacement and, therefore, my ability to be a host at all. The narrow account of hospitality seems to preclude hospitality toward any of the spaces or places I inhabit: watersheds, bioregions, and so forth. I cannot be hospitable to the very thing I inhabit.

We might try to preserve a place for the narrow view of hospitality in our relationship with nature by focusing on hospitality toward specific non-human beings. However, here we are going to run into trouble, and again for several reasons. First, as the case of Timothy Treadwell illustrates, there are any number of natural beings toward which we should not extend hospitality (in the sense of welcoming into the home). But bracketing that practical concern for the moment, there is a theoretical or philosophical problem with asserting our role as host to the more-than-human world: it asserts, without attempting to justify, an easy anthropocentric dominion over the spaces we occupy. *I* am the implaced being, the rightful owner, of the land I occupy. And, when I wish, *if* I wish, I might be hospitable to other beings I encounter there.

This is not quite parallel to the concern with distinguishing between the divine guest and the demonic monster when it comes to inter-human hospitality.[7] What Kearney calls the "wager of hospitality" is, in the case of other humans, concerned with distinguishing between the benevolent other and the predatory other; but it is not, even in a case in which we close our door to the other, a blanket assertion of superiority to all others we might potentially host. But this, too often, is precisely what happens when the other is non-human. Consider my neighborhood in Southern California. California poppies, jacaranda trees (20th century transplants), monarch butterflies, violet-crowned hummingbirds, snowy egrets, bottlenose dolphins, and grey whales are welcome neighbors. Feral peahens, Cooper's hawks, owls, skunks, coyotes, and sea lions add a bit of "color" to the environs, as long as they don't get out of hand. But we tend to be much less willing to cohabitate with bee colonies, gophers, mountain lions, black bears, or great white sharks.

We are fine with nature as a servant or vassal, and open to the idea, in certain limited cases, of nature as a guest-that-knows-its-place; but we are much less likely to be comfortable with nature-as-equal or nature-as-wild-neighbor. The problem here is that hospitality in the narrow sense, while normally a virtue, can hide a covert and unjustified assertion of privilege—in this case a facile anthropocentrism. I belong. I rule. Other things may visit or stay, but only at my pleasure.

This assertion of human priority and superiority elides the fact that those other beings also belong here. They are as autochthonous as I am, often more so. If they appear out-of-place, it is only because of the merciless and destructive excess of human dwelling and inhabitation.

For these reasons, if there is a value to the idea of "hosting Earth"—and I think there is—it is going to have to be articulated in terms of a more expansive sense of hospitality.

Hospitality as Holding Back

French philosopher Michel Serres has written a number of books that bear on this issue. In *Malfeasance: Pollution as Appropriation*, Serres reflects on how animals inhabit the places they call home.[8] They do this, he observes, by marking territory in order to appropriate it. This is generally done by depositing some form of bodily "dirt": urine, feces, and so on. "To make something its own, the body knows how to leave some personal stain: sweat on a garment, saliva or feet put into a dish, waste in space, aroma, perfume, or excrement, all of them rather hard things…"[9] These bodily markings or deposits are not merely notifications, "I was here"; they are appropriations, "this is mine." In Latin *to inhabit* (*habitare*) and *to have* (*habere*) come from the same Proto Indo-European root, *ghabh-*, indicating that to inhabit is in some way to possess, and to possess is a kind of inhabitation.[10] Serres's claim is that this kind of appropriation-through-pollution is widely distributed among different forms of life, and evident in a variety of human behaviors: interning the bodies of the dead, spreading nightsoil, burying or burning trash, scattering litter. The surest way to ensure possession is to pollute: he who spits in the soup secures his uncontested ownership of it.

However, if appropriation-by-pollution is common in the animal kingdom, humans have taken this practice to another level, because we have polluted the Earth itself.[11] We have changed the chemical composition of the atmosphere and, through it, the very climate of the planet. We have altered the pH of the ocean, making it more acidic. Microplastics are found not only in the soil and the sea. They fall from the sky in raindrops. They are found in the human bloodstream, in placentas, in breast milk. There is no place on earth free of human pollution and, therefore, no place on earth that is not under the threat of explicit or implicit human appropriation. That fact that we've marked ourselves and our children in this way is a perverse auto-appropriation in which we've colonized ourselves, largely without understanding what we are doing, and certainly without any kind of consent. Is it any wonder that we think of our current age as the Anthropocene? A new epoch in which human impact dominates, well, *everything*.[12] That domination is largely a factor not of our numbers (there are many more numerous species) or our direct physical domination (as in, for example, domestication and animal agriculture), but of the omnipresence of our waste.

Needless to say, this behavior—global appropriation, without limit—is inhospitality *par excellence*. We seem to be unwilling to exercise any restraint, to hold back in any way, unwilling to leave any space unmarked, leave any room for the wild, non-human other. Think of the cloud of trash currently orbiting our planet; we've begun to appropriate space itself. Like the greedy businessman asked how much wealth would be enough, our answer to how much we intend to appropriate is, apparently, *all of it*. Here, I think, is the key to a more expansive sense of hospitality, one that is essential in dealing with the more-than-human world, but also one with which we are failing miserably.

Fundamentally, hospitality has to do with making space for someone or something else. Even in the narrow sense of hospitality, when we invite someone into our home or to sit at our table, we can see that "making space" for the other is essential. This commitment to leaving space or making space is the essence of hospitality when applied to the Earth: holding back, exercising restraint, resisting the temptation to expand, mark, appropriate.

I'm not referring to cases in which a person might be tempted to take something that already belongs, in some relevant sense, to someone else. It is no virtue to refrain from invasion or colonialism.[13] I refer, rather, to space to which everyone—or, extending the point to include non-human animals, everything—has a valid claim: a commons, a public good or space that is freely accessible to everyone. No one owns the beach; it is a public space that anyone can access.[14] However, there is a kind of hospitality at work when people do not take up more space than they need, and when they behave in a welcoming way toward others, even though those others already have a right to the public beach. Hospitality like this operates all the time among people sharing common resources, for example, making space for others at a campground at which you arrived first, or taking care not to unnecessarily crowd others at a campground when they arrived first.

Thus, while narrow or strict hospitality is, perhaps, impossible in the commons— after all, you have as much of a claim to camping on BLM land as I do—hospitality in an extended sense is absolutely essential for the sustainable management of a commons. A commons is degraded when individuals take more than their fair share of the resource.[15] Thus, the key to the sustainable use of any commons is that people *take less than they can*. That is, they hold back.

Holding back, according to Serres, is essential to understanding and ethical behavior in a whole range of areas: ecology, theology, philosophy, science, culture, language. This is because, unchecked, things tend to follow a "general law of expansion." Like matter in a gaseous state, they spread to fill the available space. Kudzu vines expand across terrain covering it in thick, green leaves; but in so doing they block out sunlight and kill other trees and bushes in the ecosystem. Yeast will expand in wine fermentation, consuming available sugars until they quite literally die in their own waste (i.e., ethanol). Whether human impact on the Earth is more Kudzu-like or more yeast-like remains to be seen. As is whether we will be able to change our behavior and exert some restraint.

Serres argues that unchecked expansion is characteristic of naked ambition, violence, and barbarism, to which we could add various forms of authoritarianism, war, and colonialism. "Evil gets around, that is its definition: it exceeds its limits."[16]

Morality demands the renunciation of the law of expansion and begins with abstention. Our first moral duty, though not our last, is *do no evil*.[17] It is perhaps for this reason that Serres refers to wisdom as "pacified knowledge," knowledge that rejects blind fidelity to the law of expansion and the violence it entails.[18] Philosophy is not the love of power, but the love of wisdom; it is the capacity for thinking and the desire for truth decoupled from the compulsion to dominate.[19] The wise person does everything she *should* do, not everything she *can* do. A first step in this direction is to abandon the ruinous commitment to the Protagorean "man is the measure of all things." The world does not radiate from my feet like lines of latitude from the poles of the planet; I do not occupy a zero point, the navel of the world, around which the rest of reality is organized.

Holding back is what hospitality looks like on the Earth. Exercising humility and restraint. Taking just what we need rather than taking everything we can. Leaving space for others. This requires that we renounce the goal of appropriating the entire Earth for our own purposes. And it means holding back from various forms of pollution, which are the means by which we appropriate. We must resist the temptation to fill every space—physical, cognitive, linguistic—with "me" or "us." We do this by sharing the Earth with other beings, by leaving space for them, exercising restraint, and allowing some places, and the beings that inhabit them, to remain free to express themselves in their own inhuman ways.

When Irish writer John Moriarty gave up a teaching position at the University of Manitoba in 1971, it was to return to the Gaeltacht of Connemara, Ireland—initially to the remote island of Inisbofin—in order to find his "bush soul." It's not difficult to hear Irish echoes of Thoreau's desire to reconnect with his wild self.[20] One of Moriarty's explicit concerns was to reject a society and culture that shaped nature to suit itself rather than, or to the exclusion of, allowing nature to shape it.[21] If we shape the Earth, domesticate huge tracts to satisfy our wants, pollute and appropriate much of what is left, there will be no meaningful wildness left. And with no wildness, civilization and culture themselves suffer. Thoreau reminds us that "village life would stagnate if it were not for the unexplored forests and meadows which surround it. We need the tonic of wildness…"[22] And Moriarty compares Western civilization to a potted plant that has exhausted its soil; the only way to restore its health is to transplant it into wild soil so that it can continue to grow.

This does not, in my view, commit us to a nature/culture dualism that asserts humans are "unnatural" or that nature is "contaminated" by our presence. Humans are natural. We evolved here. We belong here.[23] But just as orca and elephants are different, with their own distinct goods and ways of being, so too are humans different. We inhabit the world in a seemingly unique way, that expresses its own unique goods. Human language and culture and *techne* distinguish us, in either degree or kind, from other natural beings.[24] *Homo sapiens* are not Sasquatch, and the streets of Paris are a long way from the caves we shared with our fellow creatures. Human cultures, at their best, represent a distinct good in the world; and we need not reject them in favor of some Romantic Rousseauian fantasy. However, it does not follow from the greatness of Paris that a world "all Paris" would be a good thing.

A world *all anything* is to be avoided.[25] This is why Thoreau expressed his desire for a "border life," a life with one foot in the town and the other in the forest. A life that respected both the civilized and the savage world.[26]

Ecosystems in which one form of life overwhelms all others tend not to be healthy. Imagine an island to which invasive rats are introduced. The rats have no predators, they flourish and multiply. The indigenous fauna have no defenses as the invaders raid the eggs from their nests and attack their young. The rat population explodes. However, once all the neighbors are destroyed, the ecosystem collapses, and the rats themselves have nothing to fall back on to sustain them. Except cannibalism.

It may yet become clear that humans are more murine (i.e., rat-like) than we are wont to admit; and we may have already exceeded the carrying capacity of the Earth.[27] In our industrial and post-industrial world, we habitually expand—both our numbers and our impact—beyond the ability of the ecosystem to regenerate. Humans, like other beings, seek to arrange the world for their own benefit. However, due to the magnifying power of our technology, the vigor with which we impose ourselves on the rest of the world, and the degree to which we modify it, our impact vastly exceeds that of other animals. Not since the sudden injection of oxygen into the atmosphere by primitive cyanobacteria—which transformed the planet from an oxygen-scarce biosphere to an oxygen-rich one, causing the first of Earth's mass extinctions—has there been a similar impact by one life form on a planetary scale.

An unbridled or unlimited *conatus* is madness, murderous, potentially suicidal. It is characteristic of plagues and cancer. We rightly object to such behavior in individuals and states; why then would we valorize it or even tolerate it in our species? Given the explosion of the human population and the incredible advances in our power and our impact, the only way to avoid an utterly homogenous, urbanized, humanized world by holding back, by exercising restraint and reducing the number of marks we leave on the Earth, as well as their depth and longevity. It's the backpacker's credo: leave no trace. "Leave no trace" does not mean "do not touch." We can and must leave footprints, ecological or otherwise. That, in itself, is not a problem. I am natural. So are you. Respecting nature is not a matter of avoiding contamination, or maintaining a strict, dualist wall between the human and the natural. "Leave no trace" means "do not pollute unnecessarily," which we now see means "do not appropriate." Avoid unnecessary long-lasting or permanent marks that say "me," "I was here," "this is mine."

Holding Objects and Hosting Persons: The Case for Animism

Hospitality—whether in the narrow sense or in the broader sense just described—is a *relationship*. And of course not all relationships are the same.

Gabriel Marcel observes that when we encounter things in the world, we often do so in terms of how we can analyze, grasp, or use them, whether physically or conceptually. In order to do this, we frame them as objects, things that we can hold before us for observation, quantification, and manipulation, things entirely separate from ourselves. The one exception to this seems to be other persons. True, we sometimes analyze, grasp, or make use of them for our own purposes. However, with persons there is always a sense that objective analysis misses part of what is essential, that trying to possess a person is inappropriate, perhaps impossible; and that they have their own purposes independent of their use to us. We treat people differently than we treat all the other furniture of the world because they are not entirely external to us, they are presences with which we are in relationship.[28] Presence is not something we can experience by *observing* it or holding it before us; we experience it only through *participation*, by being-with in particular ways. That is to say that the experience of presence is a mode of *being* rather than a mode of *having*. This distinction is essential. Philosopher Erazim Kohák writes:

> Whatever may seem true in the simplified human environment of the cities, [in the forest, under the starry sky] the ultimate metaphysical question stands out in profound simplicity. Shall we conceive of the world around us and of ourselves in it as *personal*, a meaningful whole… or shall we conceive of it and treat it, together with ourselves, as *impersonal*…? Is *person* or is *matter in motion* the root metaphor of thought and practice? That answered, all else follows.[29]

That's quite a strong claim. The *ultimate* metaphysical choice: persons or objects? This is the kind of question raised by other philosophers referred to as "personalists": Mounier, Scheler, Marcel, Day, Wojtyła, and others.

However, hospitality to the Earth and personalism are unlikely bedfellows. Modernity extends personhood only to human beings. Humans, and only humans, are persons. This is precisely the point for the vast majority of personalist philosophers. So, for example, Jacques Maritain writes, "Whenever we say that man is a person, we mean that he is more than a mere parcel of matter, more than an individual element in nature, such as is an atom, a blade of grass, a fly or an elephant."[30] A human, he insists, is an animal; but one unlike all other animals. Humans "superexist"; a human is "a whole, not merely a part; he is a universe unto himself."[31]

However, if this form of human exceptionalism is widespread in Western philosophy and in the modern, globalized, industrialized, urban world it has helped to shape, it is far from universal. There are many philosophical and spiritual traditions that are committed to a personal or at least animist view of the natural world. This is perhaps most notable in a number of indigenous worldviews. However, rather than risk conflating the vast range of belief systems that travel, often uncomfortably, under the rubric of "animism," let me focus on one particular case.[32]

Robin Wall Kimmerer is a botanist and ecologist. She is also an enrolled member of the Citizen Potawatomi tribe and the author of *Braiding Sweetgrass*, in which she champions the virtues of an animate view of non-human nature. Kimmerer argues that the Potawatomi commitment to animism is evident in the language used to describe the world.

> To whom does our language extend the grammar of animacy? Naturally, plants and animals are animate, but as I learn, I am discovering that the Potawatomi

understanding of what it means to be animate diverges from the list of attributes of living beings we all learned in Biology 101. In Potawatomi 101, rocks are animate, as are mountains and water and fire and places. Beings that are imbued with spirit, our sacred medicines, our songs, drums, and even stories, are all animate. The list of the inanimate seems to be smaller, filled with objects that are made by people. Of an inanimate being, like a table, we say, "*What* is it?" and we answer *Dopwen yewe.* Table it is. But of apple, we must say, "*Who* is that being?" And reply *Mshimin yawe.* Apple that being is.[33]

The point is that if we want to *treat* the Earth differently—for example, with restraint or hospitality—we have to *think* about it differently; and if we want to *think* about it differently, we are going to have to *speak* about it differently. When we view beings as inanimate, we immediately reduce them to *mere* resources, meaningless except for the usefulness to us. However, if we view a being as animate, we are obliged to think of it as having interests and goods of its own, as something more than a mere resource.

However, while such a perspective is perhaps distinctive of many indigenous worldviews, it is not unique to them. True, many "canonical" Western philosophers are clearly committed to human exceptionalism and an objectifying view of non-human nature. Descartes famously asserted that a strict ontological difference between humans and non-humans meant the former were "lords and possessors of nature."[34] Likewise, dominant, orthodox expressions of Christianity committed themselves ever more tightly to unreflective anthropocentrism and a "dominion" view of creation. Regarding the human right to possess and own external goods, Thomas Aquinas writes:

> External things can be considered in two ways. First, as regards their nature, and this is not subject to the power of man, but only to the power of God Whose mere will all things obey. Secondly, as regards their use, and in this way, man has a natural dominion over external things, because, by his reason and will, he is able to use them for his own profit, as they were made on his account: for the imperfect is always for the sake of the perfect, as stated above... It is by this argument that the Philosopher proves... that the possession of external things is natural to man. Moreover, this natural dominion of man over other creatures, which is competent to man in respect of his reason wherein God's image resides, is shown forth in man's creation (Gn. 1:26) by the words: "Let us make man to our image and likeness: and let him have dominion over the fishes of the sea," etc.[35]

Thus, Aquinas argues that while all things ultimately belong to God, humans may possess any and all of the external—that is, non-human—goods of creation and exert natural dominion over them to use them for human ends.

Nevertheless, there are, as ever, minority voices in the tradition. Despite the commonplace caricatures of Christianity as essentially anthropocentric, oppressive, and domineering, the animate and personal view of the world has not, in fact, been alien to Christianity.[36] Everyone knows St. Francis spoke of "Brother Sun" and "Sister Moon"; but he also spoke to Brother Wind, Water, Fire, and Earth. Just as Potawatomi does. John Scotus Eriugena, Giordano Bruno, Teilhard de Chardin, and a number of others articulate philosophical or theological positions that challenge the easy anthropocentrism of mainstream Christianity. The Jesuit priest Gerard Manley Hopkins tried to capture the *inscape* of things in his poetry, a coinage meant to indicate the way in which each thing "does" or "expresses" itself uniquely. In poems like "Kingfishers Catch Fire" and "Binsey Poplars," Hopkins refers to this unique expression as "selving."[37] He writes of trees "unselved" by being cut, which sounds much more like the death of a person than the harvesting of a resource. Thomas Berry, a Passionist priest, wrote that "we must say of the universe that it is a communion of *subjects*, not a collection of *objects*."[38] And contemporary theologian Mark Wallace speaks about "Christian animism."[39]

It is true that the arguments of such thinkers often run counter to what we think of as mainstream—or at least historically dominant—theology and practice. They are occasionally condemned by the authorities like Congregation for the Doctrine of Faith or the Inquisition. But they are arguably in line with the experience of a large number of practicing Christians, whose sense of the sacred and the profane—and the animate and inanimate—is rather different than that of scholars at top universities. Celtic Christianity, for example, was often characterized by a more intimate relationship with the natural world. Thomas Merton writes of discovering in these traditions a "spiritual dialogue between man and creation in which spiritual and bodily realities interweave and interlace themselves like manuscript illuminations in the Book of Kells."[40]

Kohák himself extended his account of personhood to all of reality, not only humans but animals, plants, boulders, rivers, and more.[41] On this view, the natural world is not something that we can hold before us as dead matter, meaningless except for its extrinsic value to us. Nature is, rather, a community of personal realities with whom we are in relationship and of which we are a part: "The forest with its creatures, the boulders, the entire world of nature between the earth and the sky, are not a senseless, impersonal aggregate of matter in motion. That world is intensely personal; it has a life, a rightness and an integrity of its own. There is room for humans within it, not as natural robots but as persons in a personal world."[42]

Personalism goes beyond mere animacy in important ways.[43] As Kimmerer herself observes, we already think of animals and plants as *animate*. However, we deny them *personhood*. It might be objected that this is merely a terminological difference, that when Kimmerer speaks of animacy, it seems clear from her account that the worldview she describes is personalist in the sense I am developing. However, as Kimmerer notes with respect to the "grammar of animacy," how we speak about things frames how we experience them and value them. And it is not clear that mere animacy will sufficiently change our view. At least in the modern Western world, we seem to have no problem objectifying and (ab)using beings we acknowledge to be animate, whether plant or animal. Thus, while "animacy" gets us more than objectification, I suspect that "personhood" will get us more than animacy and that hospitality is one of the things personhood makes possible or more likely.

There are, of course, predictable objections to a wider view of personhood. Western philosophy and theology have an enormous stake in various, often poorly justified, forms of human exceptionalism. A full account of where, when, and how the differences between humans and non-human nature are relevant is a task for another day. However, it's worth a moment to respond to the first and most obvious objection: personalism does not require that we endorse a flattened ontology in which there are no relevant evaluative differences between humans, horses, hedgerows, and humus. We can embrace a personal view of nature and defend the claim that there are things that are distinctive, even unique, about human persons. As Kohák writes,

> [Personhood] is not contingent on the attribution of any set of traits. Nor is the overwhelming sense of the clearing as a "society of persons," as structured by personal relations, a function of any alleged personality traits of boulders and trees. It is, far more, an acknowledgement of the truth, goodness, and unity of all beings, simply because they are, as they are, each in his own way. That is the fundamental sense of speaking of reality as personal: recognizing it as a Thou, and our relation to it as profoundly and fundamentally a moral relation, governed by the rule of respect.[44]

Moreover, a personalist view of the natural world does not mean that we can never make use of it. We are part of a complex web of life, caught up in relationships of exchange that include symbiosis and mutualism, but also predation and parasitism. Our world is full of viciously interlocking teleologies. Thus, a commitment to animacy or personalism does not immediately imply that we cannot harm or use other beings. After all, few people experience moral conflicts about the consumption of vegetables, which are clearly animate. And there are, as far as I know, no indigenous cultures of veganism.

However, a personalist view would require that when we interact with beings, we recognize that they have their own goods, interests, and value. In Kant's sense, we can never treat them *merely* as means to our ends, even when we do use them for our ends, but must also always recognize their intrinsic goodness. Reflecting on the small, local farms of County Lietrim, John McGahern writes that the farmers and their animals were "all part of the same living enterprise," and laments that, now, "everything has changed. Like fishing, [farming] has become an industry."[45]

Personalism undermines facile and remorseless anthropocentrism, reminding us that we must think in a sustained, informed, and nuanced way regarding claims of human exceptionalism, and that when we make the choice to use other beings we should do so recognizing the gravity of the decision rather than simply asserting, without attempting to justify, a right. Personalism nets us additional goods as well, including a particular kind of gratitude for the natural world. Thomas Merton wrote frequently of the incarnate goodness of the natural world, commenting on the "joy and worship the [American] Indians must have felt, and the Eucharistic rightness of it!" He asked, rhetorically, "How can one *not* feel such things" when alone in the woods.[46] The answer is obvious: our inability to appreciate the "Eucharistic right-ness" of nature is a consequence of the oft-observed and lamented "disenchantment of the world," which itself is an example of its depersonalization.

Finally, and importantly for the present inquiry, a personalist view of nature seems absolutely essential to any genuine hospitality toward the Earth or any of the myriad non-human beings with whom we share it. It seems nonsensical to speak of "hospitality" to objects, or even animate-but-impersonal beings. Absent something like personhood, hospitality is in danger of slipping into vague gesturing toward an "openness to the other." Of course, openness to the other *is* characteristic of hospitality; but it is also characteristic of tolerance, curiosity, open-mindedness, and a host of other dispositions quite distinct from hospitality. If hospitality points to something specific—whether in the narrow sense of hosting someone in one's home, or in the broader sense I've tried to articulate here—it seems like person-hood is going to be central to the relationship.

A homeowner might tolerate a tree, the roots of which are breaking up his drive-way, for any number of reasons. Perhaps the tree provides fruit that is particularly beloved by the homeowner's partner. Perhaps it provides needed shade to cool the south side of the structure during the summer. But I don't think that the openness of the homeowner to the tree is going to resemble hospitality unless and until the homeowner thinks of the tree as a kind of person: as a being that has its own goods, which it pursues in its own ways, a being that uniquely "selves," one with which I am capable of being in relationship.

I realize that for many people, the idea of personhood is a bridge too far. But, re-ally, is it so far-fetched? Many people who have pets or companion animals already treat them as persons in the sense I've outlined: as beings with a unique identity, with their own goods and ends, and with intrinsic value independent of their relationship to us. Of course, we anthropomorphize when we treat animals as *human* persons, dressing them in sweaters, imposing vegan diet, sending them to pet counselors, and so on. But when we decouple personhood and humanity, we open up the possibility of relating to our non-human companions as persons without the confusion and error of anthropomorphizing them. There is nothing in principle keeping us from making similar distinctions with other beings. The personhood of my wife is different than the personhood of my dog. I'm in no danger of conflating the two, or in any way denigrating my wife, just because I recognize the intrinsic goodness and selving of my dog. Likewise, the personhood of my dog and the personhood of the blood or-ange tree in my backyard are different, and neither is the same as the personhood of the granite boulder outside my study. In the widest sense, James Lovelock's "Gaia hypothesis" already suggests something like the selving of the Earth, and on some accounts—there are different versions of the theory—perhaps personhood.

Of course, people raised in late 20th or early 21st century industrial cultures, and the worldviews that underlie and perpetuate them, are going to have a tough time getting from the philosophical or ethical acceptance of personhood (which we can achieve via argument) to the lived and felt experience of an animate or personal world (which we cannot achieve through argument alone). There is no easy solution to this problem. However, as with other biased worldviews, we can begin by talking about things differently. As Kimmerer observes, when we speak about things differently, we begin to experience them differently. However, I don't think that simply speaking in animist or personal terms is going to do the trick in terms of hospitality. Like any relationship, our relationship with the natural world is made healthy not by *quality* time, but by *quantity* time. It's not enough to take a trip to the mountains or the woods or the ocean and speak a few words about its value, no matter how well-intended, only to retreat to a life alienated from nature. We don't uproot biases simply by talking about them; we need, in addition, to change our experience and our behavior. We need to integrate the experience of nature into our lived experience. This will include both the preservation of inhuman, wild landscapes and the integration of nature into our everyday lives, even as we dwell in suburbs or cities. Jack Turner calls this "gross contact," and contrasts it with abstract reflection. It is no surprise that many of the people who develop a personalist or animate view of nature—Erazim Kohák, Henry David Thoreau, John Moriarty, Robin Wall Kimmerer, et al.—are folks who spend time with it. Of course, gross contact is not a magic wand. Plenty of people who work the land and spend their entire lives connected to it also treat the more-than-human world as mere object or resource. But people who theorize about nature from a climate-controlled office in a megalopolis tend to miss the mark; they value or fall in love with an idea rather than a concrete reality.

* * *

Hospitality, when directed toward the more-than-human world—non-human ani-mals, plants, abiotic entities, ecological systems like watersheds or bioregions, and ultimately the Earth itself—is best understood not in terms of sharing one's home, as would be the case with other human beings, but rather as a commitment to hold-ing back. Sharing the commons. Not appropriating unnecessarily. Not polluting. Using what is necessary, with respect and gratitude, but leaving the rest. Those kinds of dispositions and attitudes are more likely to be achieved if we can bring ourselves to widen the circle of personhood so that it includes more of our fellow creatures. We need to stop behaving as if we are lords and masters of a world that belongs to us and start behaving like residents of a world to which we belong.

POIESIS OF THE EARTH: "A BLACK AND LIVING THING"[47]
Catherine Keller

is a black shambling bear
ruffling its wild back and tossing
mountains into the sea

is a black hawk circling
the burying ground circling the bones
picked clean and discarded
is a fish black blind in the belly of water
is a diamond blind in the black belly of coal

is a black and living thing
is a favorite child
of the universe
feel her rolling her hand
in its kinky hair
feel her brushing it clean

Lucille Clifton[48], "the earth is a living thing" from *The Book of Light*. Copyright © 1993 by Lucille Clifton. Reprinted with the permission of The Permissions Company, LLC on behalf of Copper Canyon Press, coppercanyonpress.org.

In an earth-grammar that pulses with succinct endlessness, Lucille Clifton (1936–2010), one of the leading American poets of recent decades, plies the path of a radical ecopoetics. Early and prophetically, in poem after poem, she roots and twists her life as a Black woman into the life, the lives, of the planet. The "earth is a living thing": and it steps forth not just as organism but as animal. One may think of recent conversations about Gaia as a "superorganism"—not a goddess, not a person-like totality, but something like "a self-regulated system."[49] But let us roll with Clifton's imagery: "A black shambling bear."

The bear's animation, its animal aliveness, conveys the sheer power of the wild. "Ruffling its wild back and tossing/mountains into the sea." Ferocious in its beauty, the image also tosses away any (invariably white) mama earth sentimentalism. Nor does such a toss serve the ends of an environmentalism that neglects the feral power of the earth itself. The shamble, the ruffle of the great black bear carries the force of monstrous quakes, shifts—the earth deterritorializing itself.[50]

Now, more than three decades after Clifton wrote the poem, we are more likely to imagine the earth tossing Manhattan than a mountain into the sea, enraged at the toxic emissions and destructive ecologies of urban civilization, impatient with our abuse of its hospitality. But maybe such eco-impatience is a human projection. Tinged with revenge, such an image might only intensify the feeling of alienation from our home—an alienation that has enabled every phase of the abuse. Perhaps the metaphor of the earth-animal's sheer power and volatility works better *without* projecting eco-vengeance onto it, without injecting that wildness with impatience and rage. Still, that bear's mountain-toss does convey an energy akin to anger. As our sick civilization presumes upon earth's hospitality, violates and abuses it, as it fails to recognize the hosting earth for what it is—why would the planet *not* become scarier? Why would the extreme storms, fires, floods not have begun so measurably to amplify?

Ruffling its back ever more wildly, the earth-bear may not be best read as vengeful—but perhaps as a host turning slowly hostile. The guests show more greed and indifference than appreciation to this inhuman host. Who is nonetheless still hosting, and evokes Jacques Derrida's neologism *hostipitality*, hospitality fused with hostility. Yet Clifton's opening image insistently avoids anthropomorphic affect projection. And the poem itself does not swerve from its sheer delight in the wild varieties of earth's beauty. Nonetheless a radically different tone, a tragic perspective, cuts through her poetry.

Clifton captures in poetry the emergent work of ecofeminism and anticipates the distinctive voice of ecowomanism. With the tradition of the enslaved ancestors she proudly claims, as Abdel Mohsen Ibrahim Hashim demonstrates, she "calls for interconnection not only with nature, but also with her African American heritage."[51]

> Being property once myself
> I have a feeling for it.
> That's why I can talk
> About environment,
> What wants to be a tree.[52]

Exposing the double commodification of earth and race, her vision finds potent resonance in the crucial extension of Black liberation theology articulated definitively by James Cone at the turn of the century in "Whose Earth Is It Anyway?" "the logic that led to slavery and segregation in the Americas, colonization and Apartheid in Africa, and the rule of white supremacy throughout the world is the same one that leads to the exploitation of animals and the ravaging of nature."[53] Clifton's warnings roll through the key decades with prophetic insistence. In an untitled poem published early in 2004 we hear:

> the air
> you have polluted
> you will breathe
> the waters
> you have poisoned
> you will drink...[54]

And so indeed human economic practices are now twisting the hospitality of the earth toward hostility. Or rather, again with a more understated and no less consequential affect:

> the patience
> of the universe
> is not without
> an end.[55]

Perhaps then all the more important—at the same time—to better mind the earth's patient cycles, to read the nonhuman vibrancy that encompasses our living and our dying. And the earth...

> is a black hawk circling
> the burying ground circling the bones
> picked clean and discarded[56]

Another antisentimental image of the planet: a black hawk circling the human burial ground. In the practice of many indigenous peoples and in Tibetan Buddhism, in the practice called "sky burials," the body after the ritual mourning gets left to the birds—particularly hawks and vultures—as nourishment. What ecological rigor is practiced in this recycling, with the picking clean of the dead body, even the skeleton, discarded but without any waste. In this nutritious cycling of death and life, the earth hosts the corpse. And in eco-reciprocity, it returns the hospitality, the body offered as food. Not much has been theologically made of this vastly pre-Christian sense of the body as host: take, eat, this is my body. Clifton tunes less to the hospitality of the dead human than to the ecology of the earth-hawk, wasting nothing. The cycle of life poses the alternative to the one-way voracity that pollutes and poisons, that does not feed but feeds upon the Earth.

> It is a black hawk circling: and the earth
> is a fish black blind in the belly of water
> is a diamond blind in the black belly of coal[57]

Blackness as earth crosses from mammal to bird to fish. And it opens into the roomy, womby expanses of sea and earth. In the elemental darkness surrounding the diamond left in the hospitable ground, lines between organic and inorganic continue to dissolve. Here in this poem free of any but the bones of humanity, the diamond is no more exploited than the coal. Or the fish. These earthlings live at home in the earth that they are. Here the reader may or may not think of the environmental damage of mining, particularly coal mining. As it happens: coal is the single biggest contributor to anthropogenic climate change. Burning coal accounts for 46 percent of carbon dioxide emissions and 72 percent of the greenhouse gas emissions from electricity. Unpoetic facts. Clifton's poem came out in 1993. Without imposing intentionality, we note that her poem came out just as—in the early 90s—through improved computer modeling and analysis of the ice ages, the consensus formed around a significantly older hypothesis: human-caused fossil fuel emissions were bringing discernible global warming. Is Clifton here warning? Or perhaps—attending. Tending.

And now suddenly materializes, as the final stanza, the most tender imaginable picture of our planet. The earth…

> is a black and living thing
> is a favorite child
> of the universe
> feel her rolling her hand
> in its kinky hair
> feel her brushing it clean[58]

The preciousness of the earth, so galactically rare, possibly unique, in its hospitality to biological life, is here imagined as a favored child, beloved of the universe.

A universe playing with her child's hair. This ritual of affectionate brushing, with hair so textured, thick, a pleasure to play with, is not unfamiliar in African American culture. Through the image of the living earth—as child, not mother—Clifton returns to planetarity the vitality of sensuous feeling, of touch.[59]

It is only in this last stanza that for this poem blackness manifests its racial signification. How then can the reader miss the reverberation between Black life and the life of the planet? Gently, succinctly, Clifton delivers a fresh metaphor of the relationality of the cosmos itself. The universe—a universe unknowable in the expanse of its dark energy, dark matter—embraces the earth in intimate care. And there is nothing casual for Clifton about the convergence of the cosmic with the racial blackness. On the intersection of racism with the economics of exploitation and environmental destruction, her vision is unfailing:

> only to keep
> his little fear
> he kills his cities
> and his trees
> even his children. Oh
> people
> white ways are the way of death
> come into the black
> and live.[60]

Come into the black and live. Whether I, and others of my pallor, will come into that black, and with other earthlings *live*—remains questionable. But at least we may hear, feel, heed the call. We may discern what deadening ways remain our own, systemically, spiritually, planetarily. And in order to face, and continue to face, the deadliness, critique needs the supplement of poetry. A poetry that is also an activation, a becoming, a creating, a making, in the first sense of *poiesis*.

So we may keep returning to the image of earth's delight in its universe-mother-home. This earth is pictured not as polluted, dirtied, but as being brushed clean by the universe. Her poem does not yet suggest brushing it clean of humans. If we could live in the imagination of our earth so hospitably held by the cosmos we might appreciate the asymmetrical but reciprocal relations of care. We might more urgently protect the earth that in asymmetrical immensity hosts us. Even as the universe hosts the earth. But how can we earthlings host our earth-host?

Perhaps only playfully, like a favored child, with no resentment of our utter dependence? But this would mean for our species to become-like-a-child-again; this species that has lost all childlike innocence or simplicity; this species with a few doing most of the exploitation, jointly wasting the matter of the earth *mater* and the mattering of most of its people. And that majority of people, particularly of color and female, will only suffer more and more from the depredation of the planet.

Nothing about this 1992 poem fades from resonance: the earth remains wild black bear and delightful black daughter. Yet as the extractions and the extinctions roll on, the impact of our species on that child has worsened. Now we find ourselves in the decade of countdown to irreversible climate catastrophe. Of raging deterritorializations that will, without radical reterritorializations that bind back to earth our systemic priorities, toss more and more of its creatures into the abyss. We are left with no reason for optimism. But perhaps we may sense a shadowy hope, a hope darker of skin and universe. In its blackness, to become earth child again cannot mean erasure of all that has been done and learned. This childlikeness does not childishly eliminate responsibility; it might though energize the ability to respond. It might empower the disempowered. It might let us face the unbearable damage. Even face the bear's rage. Responsibility—when it is an ability and not just a demand—enacts the reciprocity of responding and being responded to, of hosting and being hosted, of touching and being touched, of making and unmaking—and yet again, creating.

The last word, the swerve from the past to the possible, from power to poiesis, can only be Clifton's:

what has been
can be unmade

it is perhaps
a final chance.[61]

HOSTING EARTH: SOME BRIEF PHILOSOPHICAL AND POETIC CONSIDERATIONS
Jason M. Wirth

It is critical when we are speaking of hospitality that we become willing to host all accounts of hosting, to go beyond the habitual demand (largely reinforced, alas, by our own educations in philosophy) for the one true account of hosting, the great form or ideal of hosting, and therefore the automatic demotion of all other accounts as strangers to the "true" host. Hospitality is not simply the most persuasive account—although its philosophical apologies are compelling—but also a conversion, a change of heart, a turnaround of the mind so that it is more generous and welcoming.

That being said, it is important, in light of the theme with which we have been charged to address, to remember that we are first and foremost being "hosted." Not only are we being hosted by Richard Kearney and Melissa Fitzpatrick, but they also acknowledge a deeper and more comprehensive being hosted in their beautiful book, *Radical Hospitality: From Thought to Action*: "The challenge might be raised that in regard to nature, we are in fact the one's being hosted, which is certainly true. We are guests on this earth; and nature in turn demands to be a guest in our lives whether we like it or not."[62]

As Zen Master Dōgen (1200–1253) counseled in *Bodaisattva Shi Shōhō* (The Bodhisattva's Four Methods of Guidance, 1243), generosity to the stranger—what the Dharma calls *dāna*, gift giving—is also born of the cessation of the ego and its concomitant greed and aggression. In this sense, giving is also not taking, not consuming, not absorbing everything into oneself. Dōgen: "To leave flowers to the wind, to leave birds to the seasons, are also acts of giving."[63] The Daoist classic the *Zhuangzi* speaks of the true person who "leaves the gold in the mountains and the pearls in the sea."[64] These ancient versions of giving as not taking can also now be read as the realization that hosting the earth that hosts us includes the need for de-growth and the turn toward a sustainable stable state economy. Kearney and Fitzpatrick have their own version of this ancient virtue: "Leaving nature alone—this is, respecting nature—is itself an act of hospitality that refuses hostile acts of exploitation and manipulation. Welcoming nature into our lives and letting nature welcome us back is another step yet to be taken" (RH, 109). I would add that this takes the pernicious fantasy of perpetual economic growth and limitless consumption off the table.

Kearney and Fitzpatrick link this virtue with hospitality as biocentrism: "moral commitment to treat all living organisms (human and nonhuman) as having inherent worth" (RH, 109) as well as the prospects for the retrieval of something that was the mark of long lived, non-imperial, non-empire cultures: the "hospitable commons" (RH, 110). This is the form of the call to tend to that which tends to us, to act co-operatively with our shared enabling conditions, and to practice what the Potawatomi botanist Robin Kimmerer highlighted to great effect as "reciprocity" in her seminal and influential book, *Braiding Sweetgrass*. As she quotes Joanna Macey: "As we work to heal the earth, the earth heals us."[65]

The prospects for a "hospitable commons" are rooted in the realization that we are already being hosted. I share Kearney and Fitzpatrick's call for the retrieval of a "gift economy," which "challenges the dominant ideology of consumer capitalism" and "promises a departure from free enterprise market systems based on competition, scarcity, and greed in favor of a countercultural economy based on social interaction" (RH, 7). In another sense, however, we are already in a gift economy. In being hosted by the earth, none of us are strangers to it, and each interdependent being belongs in its own way in each place and in each time. Yet the dominant culture takes the gift that is each moment and place for granted. It is like we are at a great Christmas gathering for all sentient beings, but we are the one guest who hordes all the gifts, in essence taking them beyond the measure in which they are being given.

This can be seen in the raging climatic regime of the Anthropocene which, in a word, announces the death of the ancient Daoist and Buddhist sense of nature, ziran 自然, so of itself, coming to be from within itself, or even autopoiesis. In nature-cide, every day becomes Christmas on demand—a perversion of the gratitude that this ritual originally evoked. That fifty years after Schumacher's *Small Is Beautiful* there is still no meaningfully operative distinction between natural capital—the earth itself which we spend down as if it were income—and secondary capital (goods and services)—is confirmation enough of the Anthropocene's estrangement from the earth whose gift economy originally sustains it, at least for now.[66]

This is the strange challenge of the Anthropocene: to the earth, we are not strangers, but we have made the earth itself a stranger. And this is precisely what Marx, who, for all his brilliance, had no deep sense of the earth, could not think. Estrangement is not merely temporal (I am separated in time from myself, my species, and nature itself). It is also spatial: we walk on the earth as appropriators (even in Marx), thinking that we can take possession of what already possesses us.[67] We are the property of the earth in its abundant gift giving—in each moment and place receiving ourselves from the earth—but we do not have an appropriate relationship to it.

The profundity of Derrida's interventions into the gift notwithstanding, this is the ancient indigenous sense of reciprocity, "to bestow our own gifts in kind, to celebrate our kinship with the world" (BS, 31) as Kimmerer articulates it. Not the gift as a quid pro quo, but rather the realization that survival, let alone flourishing, is dependent on the maintaining the radical gift economy that sustains us. The hospitable commons requires the kind of "earth conversion" (Pope Francis, *Laudato si'*) in which, on the way to the Damascus of the Anthropocene, the scales fall from our eyes, and we no longer "persecute" the earth, the "poorest of the poor."

In this gift economy, the question is no longer first and foremost what we are—from this perspective, what we are is spatially estranged and the profaners of the original gift economy—but rather how we are. Now and here, how are we with the gift economy of the hospitable commons? How are we doing with the covenants that both welcome it and express our gratitude and responsibility?

In the photograph below by Nathan Wirth of a Coastal Oak in Northern California, the tree seems to stand alone, but it is not solitary. Its strength comes from rooting into its place, including the commons of soil, rains, grasses, insects, and the many other forms of life that make its life possible. It has become a philosophical fad to denigrate trees in favor of the rhizomatic. Although there is great force, as Deleuze and Guattari taught us, in rethinking the image of thought from its arborous tendencies, the problem is not with trees but with their cooption into a restrictive image of what it means to think philosophically. That is a point well-taken, but this does not mean that trees are not powerful teachers in an age of spatial alienation in which our relationship to land in its highly specific and concrete enabling conditions (forests are not deserts, for example) has become obscured in political and scientific abstraction. In this image, the tree gains strength from its enabling conditions and lifts itself to the heavens. It has no need to rule the world, but it is an image of living well within one's means in the gift economy to which it contributes. Trees, after all, are one of our hopes to reduce the amount of human caused carbon emissions during this time of climate emergency.

We are sometimes prone to the misconception that photography is passive, merely documenting the world. It pales before other artforms, which actively create new images. Although there are many reasons to avoid this position, one stands out as critical to the "eye" that manifests in Nathan Wirth's works. His images neither document—as if the beings of the earth were objects to be recorded by an impartial observer—nor create, if by that one means one brings something into the world that was not there before. Rather one sees how better to see. Although these images are captivating, it is not merely the case of what the camera sees, but rather how it does. The eye of the camera does not just see images, but also the elemental conditions—light, form, emptiness—at play in the image. It sees in such a way that it illuminates with the eye of the camera how the true Dharma eye sees. The images are very much real, but in such a way that it exposes the pretense of documentation as fantastical.

As we contemplate this tree, rooted in its place yet reaching to the heavens, it helps our eyes become more capacious and appreciative, less ideological and judgmental.

The earth, like our own face, is an image, to use the great Chinese Chan Master Linji's phrase, from before our parents were born. A single skillfully wrought image exposes the endless images of the earth. "To reach one thing," Dōgen claimed, "is to reach all things." It also accords with the late Bruno Latour's recent but powerful demand that we learn simultaneously "Attaching oneself to a particular patch of soil on the one hand" and "having access to the global world on the other."[68]

To host one stranger is already to invoke the hospitable commons, the awakening or realization to the Buddha lands in which we have always dwelled, albeit profanely, ungratefully, inhospitably. This call to reciprocity becomes decisive in a time in which the faith in a common world is in crisis, and we hardly know how to root ourselves into our host while promoting and preserving the sublimity of an earth of vast cultural, biodiverse, and bioregional complexity and variegation. This confirms Latour's rallying cry that "Saying 'We are earthbound, we are terrestrials amid terrestrials,' does not lead to the same politics as saying 'We are humans in nature.' The two are not made of the same cloth—or rather of the same mud" (DE, 86). How do we host the intimate singularity of our mud as we reach to the skies in an interdependent yet vast earth? *Humanus*, after all, has its etymological and ontological roots in humus, in the ground and soil of the earth, rather than in the heavens among the gods. We are mud-bound terrestrials amid an ecological emergency that exposes the *Unheimlichkeit* of our humus. We are no longer at home with the earth. We are such poor hosts that not only do we not reciprocate the hosting that engenders us, but we are largely unaware of it and tend to sabotage it. Yet this thin band of the particular conditions within which we evolved is also the only earth that we have. The way that a tree knows how to evoke the heavens has much to teach us.

Notes

1 Timothy Treadwell and Jewel Palovak, *Among Grizzlies* (Ballantine Books, 1999), 22, 127, and elsewhere. Also see Werner Herzog's documentary film *Grizzly Man* (Lion's Gate, 2005).
2 Of course, hospitality is also relevant to the more radically displaced—not the traveler or immigrant, but the exile or refugee, whether fleeing from war, sectarian violence, or climate disasters. As I write this, the Russian invasion of Ukraine grinds into its ninth month, and millions of Ukrainians are seeking refuge in countries near and far.
3 See my "Putting Hospitality in Its Place" in *Phenomenologies of the Stranger: Between Hostility and Hospitality*, eds. Richard Kearney and Kascha Semonovitch (New York: Fordham University Press, 2011).
4 Emma Lazarus's "The New Colossus" in *Emma Lazarus: Selected Poems and Other Writings* (New York: Broadview, 2002), 233.
5 There are, of course, grey areas; and I'm not suggesting host and guest, implacement and displacement, are clear and binary categories the way citizenship is (i.e., a person is either a citizen or she is not). A person can be more or less implaced, more or less displaced. The degree to which a person is a guest, or a potential guest, in my house varies from case to case.
6 Although, in *The Parasite* and in other, related works, Michel Serres gives a more nuanced take on parasitism. On Serres's account, parasitism is a kind of universal condition. Sometimes we are parasitic on others, other times we are the host to parasites.

Most often we are host and parasite at the same time in different ways, and in any case the flux of relationship ensures that, no pun intended, the worm turns. An alternative reflection on "hosting Earth" would embrace the role of the parasite and reflect on the give and take of the web of life. That, however, would be a different paper. See Michel Serres, *The Parasite*, trans. Lawrence R. Schehr (Minneapolis, MN: University of Minnesota Press, 2007) and his reading of La Fontaine's retelling of Aesop's account of the city rat and the country rat.

7 See Richard Kearney, *Strangers, Gods, and Monsters* (London: Routledge, 2002), Richard Kearney, *Anatheism* (New York: Columbia University Press, 2009), and Jacques Derrida, *Of Hospitality*, trans. Anne Dufourmantelle and Rachel Bowlby (Stanford, CA: Stanford University Press, 2000).

8 Michel Serres, *Malfeasance: Appropriation Through Pollution*, trans. Anne-Marie Feenberg-Dibon (Stanford, CA: Stanford University Press, 2011), 2. See also *Biogea*, trans. Randolph Burkes (Minneapolis, MN: Univocal, 2012), *The Natural Contract*, trans. Elizabeth MacArthur and William Paulson (Ann Arbor, MI: University of Michigan Press, 1995), and *Habiter* (Paris: Pommier, 2021).

9 Serres, *Malfeasance*, 2.

10 On this point, Kohák makes a critical distinction between possessing and belonging (Erazim Kohák, *The Embers and the Stars* [Chicago, IL: University of Chicago Press, 1987], 106–108).

11 "*The very growth of appropriation itself becomes what is properly human*. To be sure, animals appropriate their shelter with their dirt, but it is done *physiologically* and *locally*. *Homo* appropriates the *global physical world by his hard garbage* and… *the global human world by soft garbage*" (Serres, *Malfeasance*, 54). Ironically, because the ubiquity of soft pollution separates us from hard reality, we end up losing the very thing we first sought to appropriate in marking physical territory. We appropriate hard reality with our pollution and lose that very reality because other forms of pollution (linguistic, ideological, etc.) separate us from it.

12 Serres was prescient in this regard. Baird Callicott writes: "Just as Naess, Routley, Rolston, and I [i.e., Callicott] are widely acknowledged to be the founders of *environmental* ethics, Serres and Jamieson should be widely acknowledged to be the founders of *climate* ethics [that is, of something like 'the ethics of the Anthropocene,' insofar as climate change is arguably the marker *par excellence* of the Anthropocene]." J. Baird Callicott, *Thinking Like a Planet* (Oxford: Oxford University Press, 2014), 269. See Serres's *The Natural Contract*.

13 The very method of appropriation emphasized by Serres, pollution, is also the subject of Max Liboiron's *Pollution is Colonialism* (Durham and London: Duke University Press, 2021).

14 Here in California, the beaches are public. Which is not to say that any number of affluent—and notably inhospitable—landowners do not try to illegally claim ownership of this public good. First, they have pressured the State to allow them to privatize as much of the beach as possible, resulting in legal rulings that preserve the public nature of the beach only up to the "mean high tide line." However, not content with this partially successful NIMBYism, coast property owners go significantly further—illegally locking public access gates, lobbying local councils to restrict parking near access points, and even employing private security to attempt to intimidate beachgoers with false claims of trespass. At times, people aware of the law will take particular pleasure in visiting affluent or remote beaches simply to publicly assert the right to public access, and to irritate local landowners seeking to impede such access.

15 Garrett Hardin, "The Tragedy of the Commons" in *Science* 162 (3859): 1243–1248. The key insight here is that a rational actor will give in to the temptation to over-exploit a resource because the positive benefit of doing so accrues solely to the person who

over-exploits, while the negative impacts of the exploitation are shared equally across the commons. Thus, what keeps a commons in sustainable use is holding back, whether voluntarily or due to coercion.

16 Michel Serres, *The Troubadour of Knowledge*, trans. Sheila Faria Glaser and William Paulson (Ann Arbor, MI: University of Michigan Press, 1997), 119. "Alexander wept when he heard from Anaxarchus that there was an infinite number of worlds; and his friends asking him if any accident had befallen him, he returned this answer: 'Do you not think it a matter worthy of lamentation that when there is such a vast multitude of them, we have not yet conquered one?'" Plutarch, "On the Tranquility of the Mind" in *Plutarch's Lives*, Vol. 1, trans. John Dryden (New York: Modern Library, 2001), 140.

17 As Thoreau writes: It is not a man's duty, as a matter of course, to devote himself to the eradication of any, even the most enormous wrong; he may still properly have other concerns to engage him; but it is his duty, at least, to wash his hands of it, and, if he gives it no thought longer, not to give it practically his support", Henry David Thoreau, *Resistance to Civil Government* in *The Higher Law* (Princeton, NJ: Princeton University Press, 2004). This is not, I think, an argument for or justification for passivity. It is, rather, a recognition that my powers are limited. Therefore, while I am obligated to do what I can not to contribute to evil, and while I should probably work to ameliorate other evils that seem to me especially relevant, I cannot be singled out to eliminate all evil, or even some particular evil. We should all resist evil; but where we put our energy is up to us.

18 Serres, *Troubadour*, 135.

19 Serres, *Troubadour*, 121.

20 There is a curious parallelism between "the West" in Ireland and in the United States. In both countries, the West is where one goes to find the soul of the country, and to find oneself. However, in Ireland the West represents the past, a connection to the wild Irish culture that resisted British colonial domination. The West holds not only most of the Gaeltacht regions, but more rural and less industrialized communities, as well as the wild Atlantic coast. The West represents wildness in the United States as well, but one oriented toward the future via the frontier mythology. As in Ireland, the West is characterized by indigeneity, although in the US the dominant frontier myth is mostly, though not exclusively, colonial and anti-indigenous, whereas the Irish West explicitly valorizes indigenous Irish culture.

21 John Moriarty, *Dreamtime* (Dublin: Lilliput, 1999), *Nostos* (Dublin: Lilliput, 2001), *What the Curlew Said* (Dublin: Lilliput, 2014), and *A Hut at the Edge of the Village* (Dublin: Lilliput, 2021).

22 Henry David Thoreau, *Walden* (Princeton, NJ: Princeton University Press, 2004), 317–318. It is said that one can judge a civilization based on the way in which it treats its weakest members. Often this is framed in terms of how the civilization treats its children, or its prisoners, or the elderly, or some other human group. But in the Anthropocene, humans are dominating nature as well; and we could just as easily say that we can judge a civilization based on how it treats its land, its watersheds, its ecosystems, and the non-human beings—animals and plants—that inhabit them.

23 To pretend that we are not part of nature is a form of Gnosticism.

24 Indeed, one of those things might be holding back itself: the conscious choice to exercise restraint—that is, be hospitable—for the good of the other. Serres goes so far as to say that "humanity begins with holding back" (Serres, *Troubadour*, 117).

25 Even instances of people who "go wild"—Henry David Thoreau, John Moriarty, and others—are generally examples of people looking for a better way to *balance* their wild and civilized selves, not a rejection of civilization *tout court*. As Thoreau writes: "I found in myself, and still find, an instinct toward a higher, or, as it is named, spiritual life, as do most men, and another toward a primitive rank and savage one, and I reverence them both. I love the wild not less than the good" (Thoreau, *Walden*, 210).

26 In *Walden*, Thoreau reflects on his experiment in agriculture, during which he cultivated a few acres of unused land, sowing it with beans. "Removing the weeds, putting fresh soil about the bean stems, and encouraging this weed which I had sown, making the yellow soil express its summer thought in bean leaves and blossoms rather than in wormwood and piper and millet grass, making the earth say beans instead of grass,—this was my daily work" (Thoreau, *Walden*, 157). By what right, Thoreau wonders, do we force the Earth to express itself as beans—or, today, as sugar cane, the largest commodity crop in the world—rather than other, indigenous forms?

27 Of course, carrying capacity is malleable. How many people the Earth can support is, in a sense, a nonsensical question. The question is always: how many people can the Earth support at a given standard of living, that is, at a given rate of consumption? To which we must always add the ethical question: with what degree of economic and material inequality? "How many," "how much," and "how just" are three questions that are inextricably linked when thinking of human population.

28 "In the case of a presence, the very possibility of grasping at, of seizing, is excluded in principle... In so far as presence, as such, lies beyond the grasp of any possible prehension, one might say that it also in some sense lies beyond the grasp of any possible comprehension. A presence can, in the last analysis, only be invoked or evoked, the evocation being fundamentally and essentially magical" (Gabriel Marcel, *The Mystery of Being*, Vol. 1, trans. G.S. Fraser [London: The Harvill Press, 1951], 208).

29 Erazim Kohák, *The Embers and the Stars* (Chicago, IL: University of Chicago Press, 1987), 125. In a similar vein, poet John O'Donohue said of his experience of the Irish landscape: "Well, I think it makes a huge difference, when you wake in the morning and come out of your house, whether you believe you are walking into dead geographical location, which is used to get to a destination, or whether you are emerging out into a landscape that is just as much, if not more, alive as you, but in a totally different form, and if you go towards it with an open heart and a real, watchful reverence, that you will be absolutely amazed at what it will reveal to you" (Jon O'Donohue, "The Inner Landscape of Beauty," *On Being, with Krista Tippett*, WNYC Studios, February 28, 2008).

30 Jacques Maritian, *Christianity and Democracy: The Rights of Man and Natural Law* (Ignatius Press, 2012), 66.

31 Maritian, *Christianity and Democracy*, 66.

32 On the misuse of the term "animism," see See Agbonkhianmeghe E. Orobator, *Religion and Faith in Africa: Confessions of an Animist* (Maryknoll, NY: Orbis, 2018). Orobator argues that what colonial scholars lumped together under the catch-all of "animism" are in fact a complex set of beliefs that are not direct competitors to Christianity and Islam, but rather that the former is a bedrock or soil in which the latter took root. Consequently, African "animism"—that is, diverse African spiritual beliefs—can help us to cultivate ecological virtues and a commitment to the Earth (109) and remind us of the interdependence, communion, and solidarity of human beings with the rest of creation (109).

33 Robin Wall Kimmerer, *Braiding Sweetgrass* (Minneapolis, MN: Milkweed, 2015), 55.

34 René Descartes, *Discourse on Method* in *Selected Philosophical Writings*, trans. John Cottingham, Robert Stoothoff, and Dugald Murdoch (Cambridge: Cambridge University Press, 1988), 47.

35 Thomas Aquinas, *Summa Theologica*, II–II, Q. 66, Art. 1.

36 Whether Christian anthropocentrism is rooted in Christianity or in modernity, and the relationship between Christianity and modernity, is a hotly debated topic. Most famously, perhaps, in the oft-cited, oft-praised, and oft-reviled, Lynn White's (in)famous essay, "The Historical Roots of Our Ecological Crisis" (*Science* 155 [3767], 10 March 1967). For a contrary claim, hanging the blame on modernity rather than Christianity, see Merton, *Where the Trees Say Nothing*, 47. It seems to me that an honest Christianity cannot be anthropocentric, which is to forget our creatureliness. Christianity must be *theocentric*—whatever that might mean. Humans are creatures, not the Creator or First Cause.

37 Gerard Manley Hopkins, "Kingfishers Catch Fire" and "Binsey Poplars" in *Poetry and Prose: Gerard Manley Hopkins* (Rutland, VT: Everyman, 1998), 70 and 57. Note that in "No worst, there is none. Pitched past pitch of grief," Hopkins characterizes cliff faces as "no man fathomed," indicating perhaps that the wild selving of such places is beyond human comprehension or possession (86). We can appreciate such places, we can come to understand them in a sense and from our human perspective; but their wild, inhuman inscape will always elude our attempts to possess it. "What would the world be, once bereft / Of wet and of wildness? Let them be left, O let them be left, wildness and wet; Long live the weeds and the wilderness yet" ("Inversnaid," 69).

38 Thomas Berry, *The Great Work* (New York: Crown, 1999), 82.

39 Mark I. Wallace, *When God Was a Bird: Christianity, Animism, and the Re-Enchantment of the World* (New York: Fordham University Press, 2019).

40 Merton, *Mystics and Zen Masters*, 97

41 As to the supposed anthropocentrism of monotheisms, Kohák was a committed Christian (April 16, 2005 interview with Radio Prague International, "Professor Erazim Kohak— grateful to be home, grateful to be alive," https://english.radio.cz/professor-erazim-kohak-grateful-be-home-grateful-be-alive-8098254, Accessed 8 June 2022). Additionally, Jewish philosopher Martin Buber anticipates Kohák's more explicit personalism in his philosophy of I and Thou (Martin Buber, *I and Thou* [New York: Touchstone, 1971]).

42 Kohák, *The Embers and the Stars*, 211.

43 For similar reasons, this is why "person" also gets us further than mere *haecceitas*, the singular *thisness* of things. It is not that personalism is without problems. As I note above, we will need to carefully articulate the ways in which human persons are different than, for example, ursine persons, arboreal persons, or lithic persons. And we will need to further specific which of those differences matter, and in what ways they matter. But in terms of moral considerability, respect, and hospitality, "person" gets us more traction than alternatives like "animate" or *haecceitas*.

44 Kohák, *The Embers and the Stars*, 128.

45 John McGahern, *Love of the World* (London: Faber and Faber, 2010), 168 and 170. I do not romanticize the reality of the farms. The cattle McGahern writes about were destined for the dinner table. But small, personal farming has, or can have, a radically different character compared to large, industrial "concentrated animal feed lots" (CAFOs), on which cattle are reduced from living beings to mere commodities. No doubt the individual salmon in the mouth of a grizzly bear would have preferred a different fate. But the viciously interlocking teleologies of nature can play out in a ways that preserve a degree of respect and gratitude, as Kimmerer describes in her work, or in ways that demean and debase everyone involved.

46 Merton, *When the Trees Say Nothing*, 71.

47 This essay was originally delivered as a talk at "Hosting Earth," a conference at Boston College organized by Richard Kearney, April 2023. In that earlier form, the essay is forthcoming in *Hosting Earth: Facing the Climate Emergency*, edited by Richard Kearney, Urwa Hameed, and Peter Klapes (London: Routledge, 2024).

48 Lucille Clifton, "the earth is a living thing," in *how to carry water: Selected Poems of Lucille Clifton*, ed. and fw. Aracelis Girmay (Rochester: BOA, 2020), 143.

49 In the interval since her poem, there has developed a broadly respected, or at least non-dismissable, discourse of Earth as a certain kind of organism, or indeed "superorganism." Cf., James Lovelock: "When I talk of Gaia as a superorganism, I do not for a moment have in mind a goddess or some sentient being. I am expressing my intuition that the Earth behaves as a self-regulating system…" James Lovelock, *Gaia: The Practical Science of Planetary Medicine* (New York: Oxford University Press, 1991, 2000), 57. Cf., Bruno Latour, *Facing Gaia: Eight Lectures on the New Climatic Regime*, trans. Catherine Porter (Medford: Polity, 2017), 94ff.

50 Deterritorialization undoes the set of relations forming an established territory, and allows for reterritorialization. Deterritorialization thus works in a dissident complementarity with territorialization. See Gilles Deleuze and Felix Guattari, *Anti-Oedipus: Capitalism and Schizophrenia*. (Minneapolis: University of Minnesota Press, 1972).
51 Abdel Mohsen Ibrahim Hashim, "Grief for what is human, grief for what is not: An Ecofeminist Insight into the Poetry of Lucille Clifton," *International Journal of English and Literature* 5 (8) (October 2014): 182–193.
52 Lucille Clifton, "being property once myself," 19.
53 James Cone, "Whose Earth Is It Anyway?" *Cross Currents* 50 (1–2) (2000): 36.
54 Lucille Clifton, *Mercy* (Rochester: BOA, 2004), 72.
55 ibid., 73.
56 Lucille Clifton, "the earth is a living thing," *how to carry water*, 143
57 ibid.
58 ibid., 143.
59 And in order think, to theorize, to reclaim the civilizationally neglected sense touch, of course see Richard Kearney, *Touch: Recovering Our Most Vital Sense* (New York: Columbia University Press, 2021).
60 Lucille Clifton, "after kent state," ibid., 18.
61 Clifton, *Mercy*, 192.
62 Richard Kearney and Melissa Fitzpatrick, *Radical Hospitality: From Thought to Action* (New York: Fordham University Press, 2021), 108. Henceforth RH.
63 Citations of Dōgen are from the two-volume edition of the *Shōbōgenzō* (*Treasury of the True Dharma Eye*), ed. Kazuaki TANAHASHI (Boston, MA and London: Shambhala, 2010), 474.
64 *The Complete Works of Zhuangzi*, trans. Burton Watson (New York: Columbia University Press, 2013), 85.
65 Robin Wall Kimmerer, *Braiding Sweetgrass: Indigenous Wisdom, Scientific Knowledge, and the Teachings of Plants* (Minneapolis, MN: Milkweed Editions, 2013), 340. Henceforth BS.
66 This is not to say that this distinction remains unknown. It has plenty of well-known contemporary advocates, but so far it still fails to penetrate the walls of mainstream economic thinking.
67 This is not to imply that there are not robust ecological applications of Marx. Andreas Malm's *Fossil Capital: The Rise of Steam Power and the Roots of Global Warming* (New York: Verso, 2015) is an excellent example as is recent archival work being done on the late Marx.
68 Bruno Latour, *Down to Earth: Politics in the New Climatic Regime*, trans. Catherine Porter (Cambridge: Polity Press, 2018), 12. Henceforth DE.

Chapter 2

Fish Live in Water

John Manderson

Guest, Host, or Castaway?

Early in the voyage of the whaling ship "Pequod," in Melville's *Moby Dick*, Pip, the black cabin boy, entertains the crew by playing his "star-belled tambourine." Pip is the keeper of time. Otherwise Pip is not very useful in whaling. Alongside the "doughboy" he serves as a "shipkeeper" who tends the Pequod while the rest of the crew man the whaleboats, and pursue and kill whales. Occasionally a member of a crew is injured and a "shipkeeper" needs to fill in. Pip fills in for an oarsman with a sprained wrist on a whale boat commanded by Stubb; the second mate on the "Pequod"; a happy go lucky, successful hunter of whales.

When "Tashtego," the Martha's Vineyard Indian successfully harpoons a whale that panics and beats its tail to make its first run, Pip jumps, becomes tangled in the harpoon line and is dragged overboard. Stubb sees all this and reluctantly commands Tashtego to cut the line and free Pip along with the whale. The crew pulls Pip back onto the whaleboat and after they are through cursing him, Stubb advises:

> Stick to the boat, Pip, or by the Lord, I won't pick you up if you jump; mind that. We can't afford to lose whales by the likes of you; a whale would sell for thirty times what you would...

A few days later another pod of whales is sighted by the Pequod. Pip must substitute in again for the injured oarsman and when the next whale responds in panic to Tashtego's harpoon hitting home, Pip jumps out of the whale boat again. This time Pip isn't tangled in the line and in danger, so the whaleboat hurdles off, hitched to the fleeing whale. Pip is left, a castaway, to contemplate his aloneness in the expanding horizon of the sky and inky depths of an immense and "heartless" ocean. The oceans liquid bouys Pip up but his soul descends; which is to say his spiritual imagination descends to the depths:

> Where strange shapes of the unwarped primal world glided to and fro before his passive eyes... [And] Pip saw the multitudinous, God-omnipresent, coral insects, that out of the firmament of waters heaved the colossal orbs. He saw God's foot

DOI: 10.4324/9781003456940-4

upon the treadle of the loom, and spoke it; and therefore his shipmates called him mad. So man's insanity is heaven's sense; and wandering from all mortal reason, man comes at last to that celestial thought, which, to reason, is absurd and frantic; and weal or woe, feels then uncompromised, indifferent as his God.

After this revelation what is Pip and the Ocean he has found himself in? Can either Pip or the ocean be considered "guest" or "host"? While the ocean in this context could be considered the "lord of strangers" it can't know anything about welcoming and the ties of hospitality that bind host and guest. Pip is in fact "an accidental guest, a chance comer, a stranger"; a guest in the old English sense. But being a guest implies for both guest and host the ability to return home somewhere else so as not to wear out any welcomes. No, at first Pip is just a castaway. But after his revelation he becomes something else.

Some of Pip's shipmates think he's mad, but his madness comes from being shamanized by his revelation about the true state of nature. Pip is no longer a lowly "shipkeeper" who entertains the Pequod's crew by keeping time with his tambourine. Pip's revelation has shattered time and afterwards he becomes the high priest who officiates the mock funeral of Queequeg, the South Sea cannibal chieftain and harpooner. Ahab in fact recognizes Pip's new, mysterious powers, and in need of them, holds him close during the final few days of pursuit of the white whale.

It seems to me Pip's story and his experience as a castaway is relevant at levels, from the profane to the sacred, to the human problem of living sustainably on the earth whose carrying capacity for humans we seem to have reached and may well have exceeded. The "Castaway" chapter highlights particularly well the range of problems underlying the sustainable human use of ocean resources which are just an extreme, exaggerated example of the problems we face and the possible solutions we might employ to sustain ourselves in all the ecosystems of the biosphere.

Living in a Liquid Is Different from Living in a Gas

If we focus first on the profane, material world, which is what we scientists do, Pip's story elucidates something important and true about the limits of human observation of the oceans and the nature of the ocean itself. First, after Pip falls overboard he finds himself suspended at the surface of an immense and "heartless" ocean. Note Pip is suspended by a liquid, not the gases of the atmosphere or by solid dry land. The density and viscosity of the liquid result in buoyancy forces that allow him to defy the laws of gravity and remain at the surface of the liquid in contact with the earth's atmosphere where there is air for him to breath. Pip's body is left floating at the tropical ocean's surface on a windless and cloudless day. The sun is bright, but the sunlight is extinguished one hundred thousand times faster in the liquid of the ocean than in the gas of the atmosphere and it is completely extinguished at six hundred feet while the seabed may be twelve thousand feet below him. Pip's body and his senses can't be carried down alive to explore the wondrous depths. Only his "soul" can be.

Pip is a human observer. Most of the ocean is impenetrable to his senses including sight, our far field sensory modality. Pip's voyage into the depths of the inhospitable ocean can only be imaginative. The constraints imposed by his physiology and senses define an exceptionally limited observation window through which he can gather only fragmented glimpses of the true state of the ocean's nature. Despite the recent development of remarkable tools for measuring and simulating ocean features and processes we are still quite blind to many truths about states of the ocean's nature. Like Pip, we are terrestrial creatures, and the oceans, where life's media is a liquid instead of a gas, are not transparent to us but opaque. We rely a lot on imagination in our attempt to understand the ocean's features and processes. The ocean is a good "example" ecosystem because its extreme properties force us to wrestle with important truths about our nature and all the ecosystems we inhabit and depend upon. All natural ecosystems are complex and neither their structures nor dynamics are transparent to us. All ecosystems, even terrestrial ones, are "opaque" to human observers.

Nevertheless, if we are careful to think about our observation limits, and fill gaps in our knowledge with hypotheses based on first principles of biology and physics, instead of analogies to our experience as warm blooded, air-breathing creatures living along the interface between land and a transparent atmosphere; we can make inferences about the nature of the ocean and what might be different about living in a liquid instead of a gas.

Life's media in the ocean is matter in a different physical state. The ocean's liquid is 850 times more dense and 60 times more resistant to stretching than the air in the atmosphere. Pip sinks very slowly, because the density of his tissues is nearly the same as the density of the ocean's liquid. Resistance to stretching is viscosity and the high viscosity of seawater means that drag rather than gravity is the dominant force controlling the movements of animals and other materials. Except for things attached to the seabed, everything in the ocean, including Pip, drifts downstream. Most marine habitats are not fixed in geographic space, but dynamic in both space and time in "hydrographic space." They are, more or less, forever on the move.

There is no shortage of water in the ocean and the concentration of salts is close to the concentrations of solutes in tissues of organisms. Maintaining the balance of water and salt required by cells for metabolism is relatively easy in the ocean compared with on land where the atmosphere is dry and devoid of salts. On the other hand, the rate of heat transfer in liquid is hundreds of times faster than in air and the metabolic cost of maintaining body temperature 1 degree higher than the external environment is over 3000 times more expensive in the ocean than on land. The oxygen required to support the metabolic costs of warm bloodedness is also scarce in the ocean, so most water breathing marine organisms are "cold blooded." The metabolisms, physiologies, and behaviors of marine organisms are tightly coupled to ocean liquid while the terrestrial organisms are decoupled from the dynamics of the atmosphere to a much greater degree by gravity and physiological and behavioral adaptations. This is why the impacts of global warming on the distributions and dynamics of marine populations are larger and are occurring at rates that are an order of magnitude faster than for terrestrial populations.

Processes supporting food webs in the ocean are also profoundly different. Viscosity is important in the ocean but gravity still operates. Pip still slowly sinks. As materials, including essential nutrients, drift downstream they sink beneath well-lit surface waters where plants can photosynthesize them to create the carbohydrates, sugars, and free oxygen upon which all food webs, directly or indirectly, depend. As a result the means of primary plant production in the ocean are the forces that stir it; forces that mix up and concentrate nutrients in well-lit surface waters where plants take them up to complete photosynthesis. These stirring forces include tides produced by the motions of planets, the earth, moon, and sun; current flows driven by winds and rains produced by the earth's turbulent atmosphere. Humans can't assist in "fertilizing" the oceans as we can on land. We don't control the sublime planetary and atmospheric forces that are the physical processes that control the means of ecosystem production in the sea.

Pip's insight into the sublime nature of the ocean, its wild immensity and indifference to humans, is consistent with our scientific insights into the fundamental nature of the ocean and the forces driving the dynamics of its ecosystems. Thinking of ourselves as "hosts" of the ocean misses the mark. Humans don't manage oceans and the planetary and atmospheric forces driving ocean ecosystems and their productivities. We can only manage our own activities in opaque ocean ecosystems that we don't control and barely understand. This is true of all ecosystems in the biosphere; the ocean is just an extreme case.

Metabolic Imperatives

If we can't host the oceans, can we think of ourselves as guests? The uniquely different adaptations of marine and terrestrial organisms are the solutions to the problem of maintaining generally similar metabolic processes in a salty liquid or a mixture of gasses, two profoundly different media. Ultimately all of life's processes from the normal functioning of cells, through the dynamics of populations and communities, are fueled by the metabolic process of respiration which requires some sort of fuel; complex organic molecules that can be converted to forms of energy useful for "biological" work. Life feeds on raw materials composing the metabolisms of "others." Many organisms including humans subsist by killing and eating. Humans are unexceptional in this. That is our nature.

It's important to recognize that seafood is an important source of food for humans. Fish and other seafoods are nutritionally important to 42 percent of humans on the planet and the sole source of protein for 16 percent. At noon last Tuesday (4/18/22) there were approximately 7 billion, 900 million people on the planet who each required approximately 2000 calories per day. The human population required 17.6 trillion calories last Tuesday. About 3 billion people depended upon seafood for nutrition, and 1.3 billion of these depended on seafood as their sole source of protein. Most of the human population fully reliant on fish protein doesn't inhabit wealthy nations; they live in the developing nations. For example, a large amount of Atlantic mackerel harvested on the US east coast is shipped frozen to cities in North Africa. There it is thawed out, salted and sold in markets of cities like Cairo, Egypt. Buyers carry salted mackerel to villages where the fish are smoked and sold again. The Smoked mackerel is transported to remote rural areas that lack refrigeration. Most people eating fish live at subsistence levels without any modern conveniences.

Fish matter to humans as food. It is important to be mindful of the tradeoffs of meeting human metabolic imperatives using different food resources. I have a close fisherman friend named Bill Bright who owns a mackerel/squid trawler. During the summer of 2018 he harvested 5 million lbs of "summer" squid. Bill catches squid with a net towed close to the seabed but other species are rarely caught in his nets. With respect to bycatch, the "summer squid fishery" is clean. If you do the math, Bill's 5 million lbs of squid contained about 2 billion calories. The energy equivalent of all that squid is approximately 23,760 bushels of wheat. An acre of land can yield about 48 bushels, so about 500 acres of land are required to grow the energy equivalent of my friend Bill's catch of squid in 2018. The wheat production requires demolishing a diversity of habitats and at least displacing the organisms that use them on 500 acres land. The wheat monoculture is created and maintained by human force of tilling, watering, fertilizing, and applying pesticides each year. Bill caught 2 billion calories along the outer edge of the continental shelf where that squid productivity is supported by the stirring of the ocean by winds and tides; by the gulf-stream, its rings and its streamers. These are planetary and atmospheric forces humans don't control. The ocean does it itself.

The upshot of this is that humans depend for subsistence on food produced by weakly bounded complex ocean ecosystems driven by forces we don't control and barely understand. We aren't hosts to the oceans because we don't control the forces controlling the means of ecosystem production. The ocean isn't a "host" since it can't know a thing about hospitality. Pip knows "it's an uncompromised, indifferent god." We aren't guests either because we cannot leave and go "home" elsewhere. We are fully dependent on the ocean and its ecosystems. Our unexceptional metabolic imperative along with other necessities leave us inextricably embedded in the earth's ecosystems. Our dependency upon them is not negotiable.

Chapter 3

People with Leaves: Invitations into Ecological Connection

Jane D. Marsching

We are dreaming of a time when the land might give thanks for the people, Robin Wall Kimmerer

All images in this essay are from the series: We Imagine Together in the Forest for the Future, 2020
From a series of approximately 12 30″ x 32′ Tyvek banners

We are in a time of crisis. We are in a time of opportunity. We are in a time of despair. We are in a time when action begets hope. We are in a time of deepening awareness of the interrelatedness of all beings. We are in a time of violent, unjust extractivism. We are in a time of powerful stories about the future. We are in a time of possibility.

DOI: 10.4324/9781003456940-5

We seek ways to repair broken connections, to heal ravaged ecosystems, to reanimate a world capitalism has made into a shopping center, to reawaken animism, to cultivate attention and love, to speak for justice for oppressed bodies, to recover a livable future. In every discipline, from finance to anthropology, from engineering to botany, from medicine to faith, new ways of communicating, knowing, studying, living, breathing, and sharing are being born (sometimes reborn). Every discipline and every individual have a superpower to bring to our current crises. Artists have always imagined worlds, re-envisioned bodies, and represented the invisible. But art is also an invitation into wonder, into relationality, and into expanded awareness, all urgently needed in the Capitalocene.

The symposium Hosting the Earth asked the koan-like question: what would it look like for us to host the earth? This question topples Western understandings of the planet as a set of services that allow humans to thrive first and foremost. When we think about hosting the earth, a whole set of images opens up before us: a dinner table with chairs designed for ourselves, plates for a planet, forks for the dandelions in our yard, goblets for the molten iron in our earth's core, napkins for molecules of CO_2 ever accumulating in our atmosphere, spoons for the storm cloud pouring rain outside. The question reconfigures our humanness, reorients our hierarchies, gives us new eyes to see, opens us to an animate cosmos. Hosting the earth as a concept might fall into the realm of poetry in its fanciful reimagining of bodies and relationships. But as a way of being, it falls more aptly into indigenous ways of knowing the interwoven connections of humans to all beings, place, and time. As a question, it is an invitation into an awareness of the reciprocity between all things.

How can artists create invitations into reciprocity? How can artworks be an enchantment? Can art open a moment of wonder that arises out of sensory experience but disrupts our everyday regulated sensory gating? Can that wonder startle us into a generous love of our earth?

A farmhouse at the edge of a forest

The Fruitlands farmhouse is nestled at the bottom of a steep hill on the edge of a forest that stretches to the Nashua River. Only about 35 miles from downtown Boston in the town of Harvard, Massachusetts, in 1843 the property had only ten old apple trees, but sat on rich river valley soil. On June 1, 1843, Transcendentalist philosophers A. Bronson Alcott and Charles Lane moved their families into the farmhouse to begin a long imagined experiment in utopian community living. The utopian founders set themselves to reimagine all aspects of daily life, from diet to family structure, from the home to economies. Their aim was nothing less than a new ideal Eden, free from traditional societal restrictions, where all persons could seek their full potential, and a model on which all future society would be based. Their rules were strict; they did not eat meat or use animal labor. They dressed only in linen so as to avoid cotton produced by slave labor. Each member's day was divided between farming and spiritual or philosophical pursuits.

No one lives out his own life but lives for all, Charles Lane

Only one farmer joined the group, but this did not daunt their determination to live off the fruits of the land. Later, Louisa May Alcott, who lived at Fruitlands briefly, wrote in *Transcendental Wild Oats*: "The band of brothers began by spading garden and field; but a few days of it lessened their ardor amazingly." In other words, farming was hard work and tough to fit in amongst all the speaking engagements, writing, and philosophizing. Alcott and Lane filled an entire room of the farmhouse with books, but only one pamphlet on agriculture. As winter descended, it became too difficult to keep the farm going. After nine months, the experiment ended. But the spirit of the experiment lived on in the writing of A. Bronson Alcott and his daughter Louisa May, Charles Lane, and all the famous transcendentalists who visited the community.

Now, 170 years later, we are still dreaming about a healthy and just future. Fruitlands itself didn't last long but the dreams lasted powerfully through their words. As Donna Haraway said in 2016, 180 years after Fruitlands was founded, "It matters what stories make worlds, what worlds make stories."[1] Fruitlands was deeply committed to telling the story of a utopian future. Their community included many people who did not fit into the norms of the time (one Fruitlander was a naturalist, and another was attacked for not shaving his beard). Their ideals were based upon equality for all (as ardent abolitionists, they did not use the products of slavery including cane sugar and cotton).

In 2019 I became an artist-in-residence at Fruitlands Museum, which began with the purchase of the Fruitlands Farmhouse by Clara Endicott Sears in 1914 and now includes a Shaker House, a museum of indigenous history, and a contemporary art exhibition program. I wanted to create a public project to instigate actively imagining the future. Frustrated with the paralysis of radical, febrile, inspiring future thinking I was seeing around me, I wanted to enliven utopian dreaming as a collective design process for our future world. If we are only imagining fearful future dystopias, how can we be coproducers of a future in which we actually want to live together?

From the lens of this moment in the time of the Anthropocene when the terrifying specter of climate change is barreling down on us, we need to co-imagine a future if we want it to become a just and healthy world for all beings. I launched my time at Fruitlands just before COVID-19 radically shifted our perspective. The great pause ignited an individual, societal, economic, and political future thinking. People were prognosticating futures on a daily basis. It was the perfect time to call upon the words of earlier futurists to shed light on where we are going. Then and now, something about the crises of our times, the darkness all around us, the stark undeniable evidence of all that is broken, imperatively demands us to cast the light of hope in all directions.

I spent a year researching the lives of the Fruitlanders, their contemporaries, utopian thinkers throughout history, and the ecology of the Nashua River Valley. I thumbed my way through A. Bronson Alcott's letters, Charles Lane's essays, receipts written in Abigail Alcott's careful hand, books on vegetarianism, spiritualism, and education—all the written ephemera collected as part of the Fruitlands archive. Trying to see into the spirit of that time and understand what utopia meant to them, I filled a notebook with words, phrases, and passages from private and published writings: "The important question here arises—"what can be done?" said William A. Alcott in 1836. "The visible world is spirit outspread" and "our choices are our destiny" wrote A. Bronson Alcott in 1877. So much urgency leapt off the fragile pages. Impossible questions posed not as a detriment to imagining utopia, but instead as a provocation to radical dreaming. Utopia not as some exclusionary zone of privilege, but ever in flux. As Steve Lambert said: "Utopia is not a destination but a direction." Utopia not as a noun, but as a verb, a process of becoming, arising out of this moment, burdened by the weight of the past.

Dreaming of a time

The heart of Utopia Press is a mobile outsized letterpress contained in a backpack. Think survivalist meets Barbarella; weatherproof cobalt blue nylon canvas is paired with chartreuse suit lining and silver prom dress fabrics. In the four removable bags attached to the frame are the materials needed to print using the wood letter forms: inks, nori paste, brushes, palette knives, paper, felt, and more. The backpack was built on a ladder frame, which was handy if we needed to get up into a tree in the forest. The wooden letters hung from a Lazy Susan-inspired wheel perched on the top of the backpack frame. Lazy Susans were commonly used at another utopian community, Oneida, to make food easily accessible to all.

The backpack is six feet high and unwieldy. It is best transported by a group of people each carrying different parts. This intentional design encourages collective use, requiring a community of people to agree upon and navigate the movement of the backpack. If carried by one person through the forest paths, the image is absurd, eye catching—both out of place and well suited to the place. When not in use for participatory events, the backpack sat in the A. Bronson Alcott's library on the first floor of the Fruitlands farmhouse next to a bust of Goethe and looking out a window into a grove of pine trees.

Over the course of the three seasons at Fruitlands, we printed approximately twelve large-scale Tyvek banners. We chose inspiring quotes from earlier thinkers and coauthored our own words for the future. Printed using ink made from trees in the Fruitlands forest, the banners hung for months in the sun, rain, storms, and winds. They bore witness to the seasonal changes in the ecosystem, shifting in the cold light of winter and the soft hues of early spring. The first banner printed was the quote that begins this section by mother, botanist and professor Robin Wall Kimmerer, who is an enrolled member of the Citizen Potawatomi Nation. The phrase chosen came from her magical book: *Braiding Sweetgrass: Indigenous Wisdom, Scientific Knowledge, and the Teachings of Plants*. The 38' banner hung from an oak and a maple tree across the entryway to the path into the forest. As we walked under the banner, we could take the teaching with us into the forest, refiguring ourselves as students of the forest, learning from it how to be in community and communion.

The banners were reminiscent of bannerols held by angels in medieval etchings, of protest banners leading marches in the civil rights era and today, of billboards, and of public interventions. They were in concert with the scale of many of the trees in the forest. Over time they were visited by museum goers looking for a quiet walk in the forest, by birds curious about the huge white things, about squirrels wondering if they were edible, by moths attracted to the white at night, and so many other sylvan denizens. For much of their life, and particularly during the quiet of COVID-19, they were part of the audience of the forest far more than any human visitors.

What happens when art is made for everyone? For every being in the ecosystem? Political theorist Jane Bennett paints a picture of a world in which everything has agency. In her concept of vital materiality, she complicates the notions of life, thing, subject, and more. But more than that, her enterprise is to upend the notion that humans are supreme and all else is in service to their needs. When we look at one of the banners in the forest, the tangle of matter includes the trees, the petroleum extracted to manufacture the Tyvek, the soil microbes floating in the wind, the water bottle cap half buried in the leaf litter, the Hairy Woodpecker devouring a rotting tree trunk—an endless index. Each "thing" touches every other thing. What binds them together? Jane Bennet writes: "In a knotted world of vibrant matter, to harm one section of the web may very well be to harm oneself."[2] This is the principle that propels Utopia Press. From medieval scholar Athanasius Kirchner's secret knots to Bayo Akomolafe's fugitive spaces, from Darwin's tangled bank to adrienne maree brown's emergent strategy, we are relearning our place in this ever changing, wildly complex, interrelated experience of being on this planet.

Listen for the future

Another world is not only possible, she is on her way. On a quiet day I can hear her breathing, Arundhati Roy

At the heart of Utopia Press is a zine: SPEAKING OF THE FUTURE, a book of 99 quotes from thinkers, artists, poets, activists, philosophers, musicians, and writers who have turned their imaginations towards what is possible with a beady eye on the challenges of the present. Intended to be a commonplace book, a collection of words on a particular theme, it can be used to orient our bodies, minds, hearts, and spirits towards the future. From Thomas More to REO Speedwagon, from Sojourner Truth to Howard Zinn, the phrases are challenging, impossible, possible, and provocative.

The downloadable font used for the wood letters and the book was designed specifically for this project based on a common typeface, Futura, whose simple forms and emphasis on modernism and industrialization have become ubiquitous today. Designed by Paul Renner in 1927, Futura was intended to discard the aesthetic, and political, ideals of the past in favor of a powerful forward movement based on geometry, simplicity, and efficiency. Used on a plaque on the Apollo Lunar Modules in 1969, Futura has been the font of space travel, of exploration, and of modernism. The typeface UTOPIAPRESS combines the fundamental geometries of Futura with those of domestic structures: barns, homes, fences. Created as 11″ high wood letter forms, the font also calls to mind the declarations of nineteenth century broadsides printed with wood type.

Utopian thinkers have long used the power of rhetoric to galvanize engagement with their dreams. At the time of Fruitlands in the mid-nineteenth century, many utopian communities in the Northeast were using newly commonly available printing technologies to spread their vision. The Millerites lived with the belief that the second coming of Jesus would happen in their lifetimes on a specific day. Preacher William Miller would hold tent revivals in which large-scale linen banners attempted to map the confluence of earthly and heavenly relationships. The texts and images could be spread widely through print technologies and became part of the ephemera of broadsides, journals, leaflets circulating across the United States.

Our notion of utopia today is a pastiche of the many ideas that have been thrown out throughout history. All are incomplete. All are riddled with limits. Even Thomas More's original utopia referred to women as "the weaker sex," among other bigoted views. We can learn both from the problematic dictates of the individuals who ran utopian communities, and also from their aspirations and attempts at creating an ideal world. The original thinkers were motivated to rearrange or abandon or reframe the conventions of their time. The new conventions they proposed aggregate into an enormous archive of utopian imagination that has accrued for the five hundred years since Thomas More's *Utopia* was published. If utopia is not a fixed end goal, a place with settled conventions, but instead a journey with an ever-changing horizon, then all those ideas are ground for us. How we leap off in radical imagination is the question.

In an epilogue to her 2014 look at global neoliberalism in India, *Capitalism: A Ghost Story*, Arundhati Roy pays tribute to Occupy after its eviction from Zuccotti Park in November of 2011:

The corporate revolution will collapse if we refuse to buy what they are selling—their ideas, their version of history, their wars, their weapons, their notion of inevitability.

Remember this: We be many and they be few. They need us more than we need them.

Another world is not only possible, she is on her way. On a quiet day, I can hear her breathing.[3]

In late fall of 2020, as we were in the grip of the devastation, fear, injustice, and deprivations of the COVID-19 pandemic, these words felt revelatory once again. The last phrase in the quote was printed across five banners hung in a pine grove in the Fruitlands forest. One could sit on a bench at the edge of the path and listen to the susurration of pine needles in the wind, the insistent call of Black-capped Chickadees, the sharp retorts from the military training ground across the river, and try to listen for the future.

Collective dreaming

Utopia Press was activated by a series of events that brought groups of people into the forest to codesign a just and livable future. From the negotiation and navigation of carrying the backpack out into the woods to the choosing of a forest glade to sit together, the events had a work-like feel to them, in which everyone had a role and no one was a passive onlooker. Utopia Press is only a tool in some ways, and tools are completed by the objects and experiences they create.

So, with Utopia Press, all the individuals were invited to sit in a circle on the forest floor, first choosing a phrase from SPEAKING OF THE FUTURE to contemplate. Then they were led through an embodied exploration of the ecology of the forest through one of the five scores found in the book. Prompts to stand on

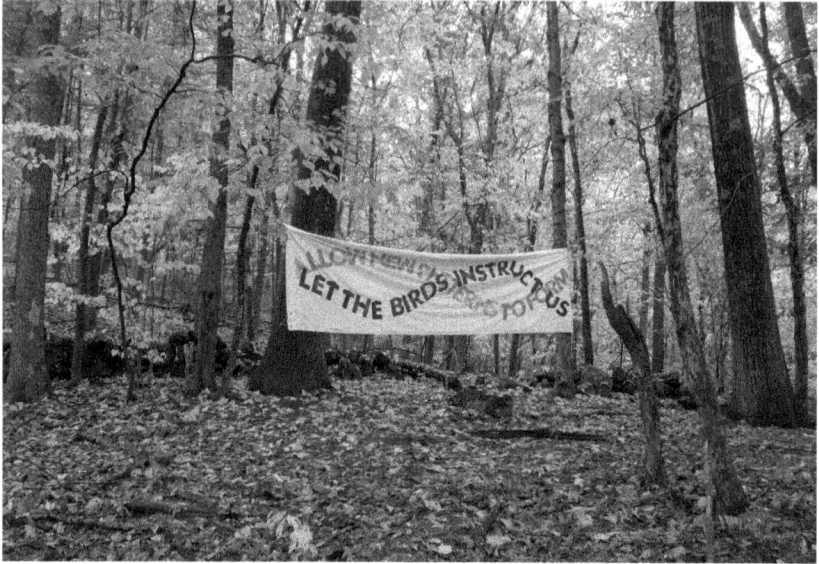

Allow new patterns to form, Let the birds instruct us, coauthored in a forest glade

your head or bury your nose in the forest floor to inhale the soil microbes into your lungs were invitations into sensory experiences of the place that shifted us away from our dependence on vision and taxonomies. Our senses were attuned anew to the world around us in unexpected and novel ways.

Then we had a conversation about what we experienced and our chosen words from utopian thinkers. People spoke about the challenges of the present: the climate crisis, racial injustice, the fears of COVID-19. They also spoke about hopes, desire, fantasies of possible futures, incomplete but earnest and germane. Out of that rich dialog came our own collectively authored dream of the future in a short phrase, our own contribution to the collective dreaming of the future.

At that point we set to making. Wood letter forms were laid out on the leaves, Tyvek paper was unrolled, and the materials for inking the letters were assembled. Everyone was invited to try their hand at printing this large-scale banner. Once completed, we used the backpack ladder to hang the banner in the forest, leaving it there for hikers and Fruitlands Museum visitors to encounter on their nature walks. The banners were vulnerable to the rain, sun, and whatever creatures were curious about them. Through the autumn rainstorms and winter blizzards, they slowly weathered, the words starting to fade, becoming more and more part of the forest.

The experiences were dreamy and communal. They were intended to shift our everyday cognition and awareness, to open us to experiences of beings in the forest that are outside of colonialist, capitalist knowledge pathways. What does it take to be in a forest and really sense our interconnectedness with all beings? Eco-theologian Thomas Berry wrote: "Among the insights attained by this meditation has been a sense of the curvature of the universe whereby all things are held together in their intimate presence to each other. This bonding is what makes the universe what it is, not a collection of disparate objects but an intimate presence of all things to each other, each thing sustained in its being by everything else."[4] This was the guiding spirit of the gatherings that were part of Utopia Press. A future in which all beings live healthy and just lives is one in which we grace each other with loving attention, the very definition of communion.

The eyes of the future

Dear future, I love you, Jane D. Marsching

Inks made from trees in the Fruitlands forest became the voice of the forest. Most of the walnut, hickory, hemlock, oak, maple, pines who gave their bark from fallen branches and trunks for the ink stood in the forest for anywhere between fifty and two hundred years. Some of them were alive to witness the life and work of Fruitlands. We can imagine them as young saplings linked to other trees through an underground mycorrhizal network that bears some similarity to the neural networks in our human brains. What are they communicating to each other? What are they sensing? How do they pay attention to the worlds they are literally making as they pull in carbon dioxide and transform it into life-giving oxygen? Who do they love? Asking these questions invites us into plant sentience in which we learn to be good kin to ones who are so beloved, generous, and essential.

On the porch of the Fruitlands farmhouse sat an open-air ink lab, which included bottles of ink, jars of acorns, piles of bark, and vessels of river water. In a solar oven in the front garden bark slowly steeped in rain or river water as its tannins were released. Foraging walks and ink making workshops invited people into the practice and ethos of ink making. Designed to be an open laboratory for sharing knowledge, the ink lab wasn't a specialized place for privileged knowledge. Post-colonial theorist Edward Said famously said "The intellectual today ought to be an amateur." He considered amateurs to be "fueled by care and affection rather than by profit and selfish, narrow specialization."[5] Following on his thinking, the Ink Lab at Fruitlands hoped to flip that premise, so that those who were beginners, or amateurs, could be invited into the space of education in the spirit of collaborative learning, shared ideas, and non-hierarchical relationships.

The collection of inks functioned as a kind of archive of the forest. In 1984 Hal Foster talked about how artists engaged archives as material, physical repositories that we encounter through our senses and bodies. Through art practice, the archive's information is released from the constrained, organized collection and reborn into the present moment. He says: "The archival impulse is made within a world whose given connections are lost."[6] What does it mean to live in a world whose given connections are lost? What connections are referred to? Who is affected and who is affecting this loss? If we shift the frame from historical object archives to ecologies, to a specific landscape, replete with plants, animals, weather, etc., it seems logical to apply the same concern about living within a world whose connections are lost. Artists can reweave those broken connections through narrative, imagination, analogy, collage, and other strategies.

The tree inks in Utopia Press served to focus our attention on those living beings in the forest, to see them as unique beings, to imagine their voice and what they might say, and to in some small way begin to mend our broken web of connections. We have all become increasingly aware of and deeply critical of our human centeredness. We are developing new language and new ways of knowing our kin. Before we can know trees, we need to observe them, listen to them, be with them. Knowing is not captured by a litany of Latin names or an understanding of ecosystem services. Knowing begins with remembering and imagination as so beautifully captured by Robin Wall Kimmerer:

> Singing whales, talking trees, dancing bees, birds who make art, fish who navigate, plants who learn and remember. We are surrounded by intelligences other than our own, by feathered people and people with leaves. But we've forgotten. There are many forces arrayed to help us forget—even the language we speak… "It" robs a person of selfhood and kinship, reducing a person to a thing.[7]

One of the inks used often in Utopia Press banners was made from hickory nuts found on the forest floor in the fall. Hickory trees are very slow to grow; a typical Hickory tree takes forty or fifty years to begin to produce nuts. The nuts have been a rich food source for humans and more-than-humans for eons. Their place in folklore is rooted in an experience of them. They have been used to augur the severity of winter: the thicker the shell, the harder the winter to come. It is said "plant your corn when the hickory buds are as big as a crow's beak," because Hickory is slow to bud out and usually appear after the last frost of spring. All of these stories speak to Hickory's experience of time: slow, patient, in tune with the seasons, which early farmers learned from in order to survive. So the Hickories in the Fruitlands forest have been patiently watching the life of the forest over decades and centuries. Who is better suited to think about the future? Terry Tempest Williams said: "The eyes of the future are looking back at us and they are praying for us to see beyond our own time."[8] Can we imagine and thus emulate the relational, generous, slow presence of trees and create a beautiful world to hold the tender threads of our interconnected planet?

Notes

1 Donna Haraway, *Staying with the Trouble: Making Kin in the Chthulucene* (Durham and London: Duke University Press, 2016).
2 Jane Bennett, *Vibrant Matter: A Political Ecology of Things* (Durham and London: Duke University Press, 2010), 1–19.
3 Arundhhati Roy, *Capitalism: A Ghost Story* (Chicago, IL: Haymarket Books, 2014).
4 Brian Swimme and Thomas Berry, *The Universe Story: From the Primordial Flaring Forth to the Ecozoic Era—A Celebration of the Unfolding of the Cosmos* (New York: HarperOne, 1994), 251.
5 Edward Said, "Professionals and Amateurs," in *Representations of the Intellectual* (New York: Vintage Books, 1996), 73–83.
6 Hal Foster, "The Archival Impulse," *October* vol. 110 (Autumn, 2004).
7 Robin Wall Kimmerer, "Nature Needs a New Pronoun: To Stop the Age of Extinction, Let's Start by Ditching 'It'," *Yes Magazine*, March 30, 2015, www.yesmagazine.org/issue/together-earth/2015/03/30/alternative-grammar-a-new-language-of-kinship
8 Terry Tempest Williams, *Refuge: An Unnatural History of Family and Place* (New York: Penguin Random House, 1991).

Chapter 4

Salvaging Islands

Fanny Howe

Ireland has a long history of island stories drawn from the moody sea and sky on the edge of the Atlantic Ocean. Here fishermen and mules and hard worked girls stood for a world that functioned on necessity. From Peig Sayers and Tomás Ó Criomhthain (*The Island Man*) to Liam O'Flaherty (*Skerret*) and Emma Donoghue (*Haven*), from the Blaskets and Skelligs to Aran and Achill, Ireland's islands were prototypes of "the little life" and they were all prophetic of their own disappearance. Such life was blessed and cursed by a plenty of water and stone, its people limited in strength and endurance.

Richard Kearney's third novel, *Salvage*, is in the lineage of this Irish literature of islands. Brigid's Island, also called Rabbit Island, lies in a cluster of other islands off the west coast of Ireland. For a reader in the 21st century the story of Maeve O'Sullivan is a familiar one. Set on an island, one that is barely surviving, the story shows this ancient world in turmoil. Maeve lives in two worlds that cannot be reconciled. We are not just reading about a lost civilization or imaginary kingdom, but about our own time. This time. With aerial bombing and tanks. Maeve stands on the edge of all that. She has always been happy on the Island with its magnificent varieties of plants, fish, birds, weather. The reader shares her happiness from the first page when the island is illuminated by the eye of the writer.

In this novel we see the natural world, currents like the ocean and the wind, and changes in temper that we cannot comprehend, although we sense how familiar they are. Now we have "salvage" as a mushroom (*fás aon oíche*), overnight growth and precious food: the fragments we can carry away hastily when evicted from childhood. How can hidden fruits of the earth contain experience, store it away in delight or terror, poison or cure. The air is still, transparent, hollow, and yet it keeps Maeve and her fellow islanders alive. They inhabit a world that cannot remain stable, cannot be tamed on land or sea. A world of horror and wonder, brinking on magical transformation while remaining solidly real.

This book has the fragility of a breath taken in the hope that all will be recalled, that a fire will blaze again. The breath in the word and the air are on all sides of it, catching the fragility of words whipping from mouth to ear and back. Natural things burst into air, the times we live in, their invisibility. All are contained here:

DOI: 10.4324/9781003456940-6

currents, electric storms, the appearance of a friend and complete disappearance too. Our isolation imperiled by the arrival of others.

Philosopher Richard Kearney tells this story in the lineage of Irish island stories, so many so marvelous with fish and fishermen, animals and phantoms, nets and sea wracks, potatoes and plants, and herbs to make living easier. Children love this literature as it comforts and teaches them, shocks and surprises. Movie makers too if films from *Man of Aran* to *Banshees of Inisherin* are anything to go by. Kearney grew up in the country he is describing, and this has produced in him, and through him, a kind of reverse DNA, whereby he grew into the body of soil and sky, and describes in his own flesh what he knows and is, creating a chemical confluence between the landscape and a growing being.

The story here told is bursting with intimate contact, with familiarity with plants and sea creatures that seem to grow from the writer's pen as the origins of life might do. Mushrooms, yes, as nourishment and healing. *Púcaí*, the islanders call them. Shapeshifters and antidotes to grief and pain. Stirrings of earth and imagination. For this is a story about belief and its fragility. Like the Irish language these islanders still speak, belief is informed by generations of elemental existence. The words that island people spoke for generations are not separate from the plants they grew and the sounds in the air. Language changes naturally over time, passed around, savored and spat out, some of it preserved in books and prayers, all the rest washed away.

Brigid was a saint after being a goddess. Later and sooner. She was an ordinary Irish woman with enormous and unusual skills. Now she is a name for a place where there is living water and a well to contain it. Brigid's Is land—*Oileán Bhríde* in the Gaelic of her inhabitants. And this gem of a book is another such container: a holding place for hope.

The Irish saint (recently erased from the Catholic lexicon, before being restored as matron of the land) has deep roots in the philosophy that Kearney has been developing for decades—anatheism, anacarnation, God after God, theopoetics, the sacred in the profane, Roots growing right along the furrow line where theology and philosophy have come to touch at last. Here that line is revealed in the love story between an island girl and a city boy on the south west coast of Ireland, where St. Brigid is still remembered in rushes and streams, stones and flowers. She was first a goddess and then fell to the status of Saint; and what she was always known for are a few scraps of lore. Here in this story, Kearney manages to turn the saint into a human girl, not a memory, where words become wood to illuminate and to warm.

Maeve is a born healer and has the Irish language born in her too, which gives her capacities that are awakened by her voice and hands. This is no more unlikely than DNA traces lasting for eons in a cold cave of ice. Traces of sound and stones and yellow petals, the same as songs in the impressions they leave and newly arouse. This is a gentle and erotic story thrown into this rough world, the 20th century only just past, and missiles and drones replacing hawks and herons. Healing is the wish behind the tale of Maeve Sullivan. And the seed of passion between herself and her loved ones. Here, the writing itself is the healing of both writer and reader. What needs to be unearthed is what Kearney never mentions. Only little cuts, blisters and bruises, and heartbreak. The sinister illnesses that come with progress and industry go unmentioned. This is a story about what is given, not made. The author's optimism is resurrectionist: A person can be made whole. The way is in ourselves, nature, and attention. God is not named, and love is only the ordinary kind, not an otherworldly imperative. So there is much to be learned from this island tale that tells us how fruitful "the little way" can be. This is another great book of intelligent religious feeling to come out of Ireland.

Note: A modified version of this essay appeared in *The Harvard Review*, Fall, 2023

Part III

Psychologies of the Earth

Chapter 5

Bringing It Back to Nature

James Morley, Sean McGrath, Edward Casey,
and Marjolein Oele in conversation with
Matthew Clemente

Psychologies of the Earth

When Freud needs an image for the development of the ego, he goes to the undifferentiated microbes at the origins of life. When he wants to illustrate the mystery of the erotic impulse, he refers us to the merging of germ-cells which furthers and complexifies existence. To show the tragic nature of human sexuality, he brings us back to Aristophanes' metaphor of man as a sorb-apple ever seeking his other half. The ego is like a horseman on the back of the id, but the untamable id leads it where it wants to go. Children are like our animal brethren, and of course we are all big children. The mind is nothing like a city, which is "*a priori* unsuited for a comparison… with a mental organism"; rather, it is better likened to "the body of an animal or a human being." Life itself is an image of "a force of whose nature we can form no conception" which awakened life in inanimate matter so many millennia ago. All of this is to say that human existence is more nearly related to the earth outside us than it is to the mad workings of civilized life. We are bodily creatures, human animals whose psyches are formed by and fashioned after the natural world, not products of human design like the artificial intelligence we are told will replace us soon enough.

Contributors to the section that follows start with the basic insight that man is not a machine but formed from the clay (or mud) of the earth and thus indebted and beholden to it. They take up this notion in dialogue with me, their humble interlocutor, whose job it was to ask them to illumine us with their thoughts on the murkiest of matters—the psyche's relation to the earth, how we both consciously and unconsciously play guest and host to our alien home. From Marjolein Oele's reflections on the dirt at our feet to Sean McGrath's speculations on the future of our species, from James Morely's humanistic ecology to Donna Orange's unearthing of our most inhuman practices, from Ed Casey's artistic hospitality to my own aesthetical musings, these dialogues are rife with insight, openness, imagination, and hope. So let me welcome you reader, stranger, friend. Let me welcome you to the work of hosting that otherness we call "Earth." I hope you enjoy the conversation.

DOI: 10.4324/9781003456940-8

A Dialogue with Marjolein Oele

Matthew Clemente: One of the topics that comes up in your recent book *E-Co-Affectivity: Exploring Pathos at Life's Material Interfaces* (2020), is this contrast that you draw between soil and earth. You start with this very moving image of burial—human burial in soil—which illustrates how soil is always changing and how, from one burial to the next, even if it's in the same cemetery on the same day, the soil is never the same. Soil is something that both hosts us—we're welcomed into soil—and we take soil with us everywhere we go. We're always eating food that comes from the soil and we're kind of living in soil. I'm curious about your thoughts on that reciprocity in relation to your work.

Marjolein Oele: My book, *E-Co-Affectivity*, is trying to do something new insofar as it tries to address the *eco*—the milieu and commonality—of our affectivity. And by "affectivity" I mean that in the broadest sense of the word, as the Greeks—in particular Aristotle—have used it. That is, suffering, undergoing qualitative change, illness and so on. My argument is that we do well as humans not just to focus on our affects and emotions, but to think beyond our own human abilities to be effective. For that reason, I try to dig into, for instance, the topic of soil. Or skin. Or the very materiality of the placenta. Because I think that it is through understanding the biological existence of our lives that we can get a better sense of our "e-co-affectivity" and how we have come to be who we are. In this regard, my book is about *becoming*, and not about being. So that already indicates a radical departure from the more traditional way of looking at our location in terms of earth, which sees us as solitary agents standing on the earth. I am seeking to engage and think beyond the more Messianic and provincial mindset associated with "earth" in contrast to the "messy" entanglements of such things as soil. Soil, for me, functions as a figurative image insofar as soil, for many of us, is a place of consolation where we bury those who are close to us. At the same time, it is also the place where we build, eat, live, and participate. We are part of the soil. So I attempt to use both material interface and the figure of soil to combat this notion of thinking about who it is that hosts—or is being hosted—and thereby I seek to displace the human and its so-called "mastery" of the earth. If we are really thinking seriously along the lines of soil—along its interfaces and pores—we might arrive at clarity about a future that is post-human. So, in many ways, though I am very receptive to your question, I would like to displace the question of hosting and being hosted as humans and step beyond the human being as we know ourselves and, instead, understand how concrete beings are being produced every day ontogenetically from the soil—from this interface—and how, if you are following along the production of this interface, we are changing as we become new beings.

MC: Along those lines, your final chapter is all about how this thinking, in terms of soil, gets us beyond the Anthropocene. Might you be able to unpack that a bit—what comes after the human? When we begin to see the world in terms of soil and start to really think this way, what comes next?

MO: My last chapter holds suggestions for the future. We often don't take the time to think beyond the human as we currently have it. There are a few characteristics of soil that give us an indicator about what new beings can become. If we think about how long soil takes to really come to fruition, then my work is an engagement with the production or generation of a post-human that takes time. Percolation—and slowing down—is crucial. Though, of course, I do not wish to undermine the very urgency of the moment we are living in. I use the example of how only now new generations of farmers can forgo tilling or employ more sustainable methods after the previous generation has passed away. In other words, our *habitus*—our way of being and our way of dealing with the soil—takes time to evolve and takes the passing off of systemic and familial knowledge to allow for transition.

MC: In your book, you go back to Plato and Aristotle. I was thinking about the first lines of the *Phaedrus*—when Socrates asks Phaedrus where he's coming from and where he's going—as I was reading your book. Is that the sort of situatedness in the present that looks back at where we come from and forward at where we're going that you're calling for with this notion of *becoming*? I wonder, as we think through the political crisis that you think also impacts and effects where we're situated now, how your concept of *becoming* and displacing our *habitus* will lead to where we're going. What opportunities are we opening towards?

MO: I think of Michel Serres' *Times of Crisis: What the Financial Crisis Revealed and How to Reinvent Our Lives and Future*. Serres analyzes crises in multiple ways. One of them is through thinking through the crisis as the body faces it in a serious moment, where either one dies or the organism reinvents itself in a new kind of way. We think of people in crisis—in ICUs for instance—and think, also, about the economic crisis—enormous poverty—and identify something there that asks us to allow access to the urgency of the question and at the same time to think about what might come out of this. There is a hope that Serres articulates: we can be reborn in a different kind of way. The crisis can be a way of being reborn. And can allow us to think through what got us into this crisis in the first place. What are the fault lines underneath this crisis? We must analyze the deeper tectonic shifts that have taken place, and, at the same time, rethink the way forward. I strongly believe, with Serres, that one of the ways forward is to engage in a new interdisciplinary intellectual direction. I urge philosophy to think through the broader fault lines of the world and do so in an interdisciplinary fashion, wherein we think along the lines of science, art and the humanities to understand what has happened to our world.

MC: Along these lines, I think in many ways it is embodying a type of philosophy that brings us down to the level of the mineral. Philosophy often takes the "God's eye view." But your book brings us down to the mineral-level—you consider bird feathers and placenta—and asks us to take on the materiality of the world and consider the things that often go unnoticed. Soil is such a great example of this—it's so central to our being—yet it's so often forgotten by us. I'm wondering if this movement in philosophy, which you're helping to bring about, could fairly be described as a shift from the macro to the micro?

MO: Yes, indeed, I am trying to turn to the matter itself. I have always been an admirer of the philosophical anthropology of Helmuth Plessner, for instance, and his attempt to think through different tiers and complexities of matter, without making it all part of one flat, horizontal surface. So, in a similar fashion, I admire the new materialists, but at the same time I want to combat this notion that everything thereby comes to be on the same horizontal level and that everything becomes flattened out. I really want to think through the complexities of the different tiers of life and non-life. I do want to delve deeply into this. For instance, something that I was really excited about was my research into bird feathers, whereby I found out how different sorts of trauma take up physical space in bird feathers. The locations of trauma are these spaces in bird feathers. But I also found out, through that research, that an animal can somehow relocate the places of trauma to wherever it is least harmful to its flight, for instance. I was able to draw conclusions—or philosophical implications—from this research. I started with the micro, and was able, in this research, to pan out to the macro, and to learn something about the complex biological activity or affectivity of an animal. There is indeed a certain beauty in the fact that a body can relocate trauma. Life knows how to negotiate with its environment. And we as humans can learn from that. We can learn from that scientific research, and as philosophers, we can very carefully, and without drawing big conclusions, learn about the materiality of our world. For me, the micro and the macro are certainly connected, but we must of course be careful before we slap a macro idea onto a micro topic. Though it certainly is possible. Another example that I have become interested in is analysis of the placenta. In my chapter on the placenta, I articulate the biology of the process as it is oftentimes unfolding. We would do well to understand what it is like to be hosted both as a possible future being coming into existence and as a possible, future mother coming into existence. I tried to do justice to this very phenomenon of being hosted by an organ that has truly magical, interesting material. Again, the micro is prominently there, and there are ideas associated with that, but one has to be super careful to rethink the implications that might arise.

MC: In relation to the topic of the placenta, one idea that you develop that I think is worth unpacking is the idea of the "in between." In relation to plants, you talk about the middle voice. I found both of these topics to be very interesting and hope you might discuss each.

MO: For me, the interface—the in between—where things emerge is most interesting. For instance, in my chapter on skin, I attempt to articulate how we become who we are in part due to the emergence of this tactile interface called the skin. Oftentimes, when we think about humans—as *homo sapiens*—we think about our brain. But in many ways, we have become who we are also through sensitivity, through the softness, through the vulnerability of our skin. So, the way our skin has evolved (and, for instance, has lost hair over evolutionary development) indicates how we have become this sapient, sentient being that we are. So often it is in the interface between an organism as it is emerging and its environment, and various factors thereof, that we become who we are as human beings. In this entanglement, we have become—and are constantly still becoming—beings. Through the surface—the interface—we find depth. This is where we find the complexity of life. And what we find there is not individual beings, but always co-affected *beings*. This is, I think, helpful as a model to make us both aware of our journey and of the possibilities of a future that is different and possible and not only disastrous.

MC: We are, in living with an awareness of the in-between, always both in the state of becoming ourselves (being affected by that which we live, we experience, we touch, we move), but also affecting everything around us. Everything is in this state of becoming for you, such that we are always affecting and being affected through the medium in which we move.

MO: That is right. And I think in that regard my training as a medical doctor has made this truth clearer to me. From the inside out—from our bowels out to our skin—we are always constantly becoming. We are always to some degree delivered over through this organism, or through this being, of which we are part. That is not to say that I do not want us to escape ethical responsibility, but real ethical responsibility, in my view, must be formulated not as something that we as human beings need to form on our own terms, but again, as something that emerges within our interfaces. I am often asked about what sort of ethics I'm thinking about. But I refuse this question, because this question is too focused on the human currently conceived as a solipsistic being. Instead, I argue, the human has to be conceived as the post-human, emerging from out of our entangled existence, in a time, a milieu, a place yet to come.

A Dialogue with Sean McGrath

Matthew Clemente: I'd like to begin by hearing some of your initial thoughts on ecological hospitality and the reciprocity that we share with the earth, as both hosts and guests of the earth.

Sean McGrath: It is crucial to consider ourselves guests on the earth. We know that the earth has preceded us by several billion years. We are late-comers and should not take our presence for granted nor assume that the "Goldilocks" conditions of the planet will always be there for us. We are all beginning to awaken to the full significance of the Copernican moment, recognizing that the destiny of the earth is not coincident with human interests. This is no doubt a humiliating moment and an occasion to de-anthropocentrize our discourse. In some respects, mainstream culture is at last catching up with the environmental movement. The critique of anthropocentrism has been the leitmotif of environmental thought from the middle of the twentieth century until Timothy Morton's work. Nevertheless, it may be that this talk is beginning to become stagnant. It may be the case that one of the reasons that the political will for making the ecological change that is scientifically demanded of us is so lacking is because this tired critique of anthropocentrism is somehow missing its target. I, for one, believe that we are not problematizing the situation correctly when we regard ourselves as merely one species among many, one population of creatures that need to be limited in numbers because its evolutionary success has become environmentally problematic. There's something about the human that is being missed when we think about it in these terms. I am an environmentalist, and yet I remain a defender of human difference. Environmental discourse makes no sense if humans are not destined somehow to govern the earth. The heavy sense of human responsibility for climate change, which is condensed into the symbol of the Anthropocene, makes no sense if we are simply one species among many. The fact that we are now—with anthropogenic climate change no longer easily denied—responsible for the fate of the planet indicates that there's something different about human beings. Something has gone wrong in humans which has led us to the particular predicament that we are in. The solution is not to deny the human difference from other species. Rather, we ought to go deeper into the human difference and consider how the human has become a sort of toxin, a poison, and a curse upon the earth and ask how the situation can be transformed.

MC: This touches on the heart of the work you're doing with your non-profit, For a New Earth (FANE), which is trying to assess why there is such indifference among human beings. Part of what leads to that is this notion that human beings are not somehow unique creatures and uniquely responsible and capable of changing the trends and addressing this crisis. It reminds me a bit of the line from Hölderlin, which, of course Heidegger quotes: "where the danger is, there also grows the saving power." By changing our gaze to see that human beings are uniquely responsible, we also see that there is something that is unique about humans that will also allow for change. What is it that you believe this change in rhetoric, perspective, and emphasis will allow for? What will we gain by becoming more anthropocentric in addressing this crisis?

SMG: The Hölderlin quote is indeed the leitmotif of my environmental work. We are nature that has become conscious of itself and, at the same time, conscious of our ecological guilt. Of course, every creature has its own agenda, but our unique destiny is to be the mirror of nature, and in us, in our moral-ecological self-consciousness, the will of nature becomes free. And this is precisely what is happening right now. The polar bear does not bear the responsibility for correcting things. I recognize that this view is characteristically Western and, more specifically, Abrahamic. So what? That is no critique unless we have already decided against the Abrahamic traditions, which include the religions of half of all living human beings. We need all hands on deck at this late hour in the fading light of the Holocene. People are going to have to come at this from a variety of cultural perspectives, particularly from a variety of religious perspectives. I am interested in developing the ecological potential of the Western Christian symbolic. I confess openly that I speak from a theocentric perspective. According to a Christian theological perspective, the problem lies in the fact that we have put ourselves in the place of God and have made nature *our* other when nature, from its original conception as *natura* in the Latin West, is the other of *God*, not the other of the human. Once the triadic relation of nature/ human/divine is restored, I believe we will achieve a more earth-centered Christian spirituality.

On a more trans-cultural note, we must, above all, address the failure of political will. We have had a half-century of ecology and climatology, and there is still no will to change. COP26 was a devastating reminder that we are hurtling headlong toward the 4-degree mark. We couldn't be better informed about it; grade-school children can explain the Greenhouse Effect. Climatology is the greatest collaborative scientific achievement in the history of humanity. And still, there is no will. So clearly, this is not a problem of communicating certain facts. It is not a problem of education and the dissemination of science. It is a problem of politics—and, indeed, of religion.

MC: I wonder whether or not this mode of rethinking the human—our place in the world—is already latent within these traditions. Where do you see the touchpoints where we can retrieve these ideas? Or is it, rather, the job of contemporary human beings to think things anew? Where can we retrieve these notions that you want to draw forth? Or should we be creating a new religious ethos?

SMG: I think that the emphasis in environmentalism on what is to come, on the strange, unimaginable newness of the ecological civilization that we're struggling to build, is a crucial motif for Christians. It represents a secular return to the eschatological core of the Gospel. Here's a place where the Christian tradition has something important to offer. But to make a contribution, Christianity needs to recover a genuinely Pauline eschatology. Christians routinely forget the future-oriented nature of Christianity. Paul's Christianity is not a religion of the past; it is not merely the commemoration of an event. It is an anticipation of something to come. We are in the midst of a movement towards something unprecedented. Christ is the midpoint of a three-act eschatological drama. The third act is still to come. When Christian thinking turns to ecology, it finds an abundance of resources to help it recover its eschatological edge. With regard to the question of the human being in a transhumanist age of dawning AI, before we assume that our technology threatens us in our human essence, we should admit that we don't yet know what the human being is, for the fullness of Christ-nature has not yet been revealed. "For now we see through a glass, darkly; but then face to face: now I know in part; but then shall I know even as also I am known" (1 Cor 13). Jesus Christ crucified and risen 20 centuries ago is the clue to the riddle of what a human being is. But the fullness of what the human is has not yet been revealed because the fullness of redeemed humanity does not yet exist. One of the gifts of climatology to theology is that it has given us new ways for thinking about the deep human future. FANE has spent much of the past eight years gathering scientists together with scholars of the humanities in spaces of public debate and exchange. We have staged public interventions in National Parks, at hydro-electric dams and in pubs in Newfoundland in order to bring the wealth of ecological knowledge out of the university and into the marketplace. An ocean chemist at one of these meetings demonstrated that we have solid bio-chemical reasons to believe that 10,000 years from now, all of the CO_2 that has been released during the industrial age will have been absorbed into the ocean. After a few hundred years of barren, acidified oceans following that absorption, the seas will calcify and likely teem once again with life. Ten thousand years seems an extremely distant point in time. But let's think about human history from the perspective of the deep past. Ten thousand years is 5 percent of 200,000—that is, about 5 percent of the age of modern humans. We have had about 10,000 years of civili-zation (the rough age of the Holocene). If we consider the whole of human history as a century, 10,000 years is about five years of that century. So as we see, we are new at this civilization game. Rather than being the "late-comers," as Heidegger put it, we are the "newcomers." We are the primitives of a new civilization that is still to come. Such a thought, firmly rooted in natural science, has more chance of engaging both the imagination and the politics of our time than all the gloomy talk of the world without us. For we now dare to speak of a deep future, not only for the planet but also for the human. Why would we be interested in a future without us? In light of the 10,000 years needed to correct the pollution produced by the fossil fuel age, we need to think about what we want for *our* future. The planet, with or without us, is fine. The planet can withstand all kinds of CO_2 and pollution. We've had way more CO_2 in the atmosphere before. What really is at stake is us. Can we have a future on the planet? Should we have a future on the planet? And if so, what kind of future?

MC: What really resonates with me here is that the apocalyptic talk around the ecological crisis is this notion that we are destroying the planet. And, as you say, the planet will survive. It is *we* who will not survive. That is what the stakes look like. You seem, in your work, to be encoun-tering—and contending with—both our shared human apathy and also those who are most interested in making a change. I wonder where you think we can redirect the conversation. Our time here on earth has been short, as you've articulated. But what do you say when you talk with someone who's used to thinking about the crisis in terms of human obliteration? Someone who envisions a future sans humans on earth? How can we return to a more anthropocentric conversation?

SMG: When we talk about saving civilization rather than saving the planet, we are forced to consider our values and attachments. What is it about our civilization that we would like to carry forward? What is worth holding on to? What forms of *human* life do we find to be worth protecting and preserving? Human beings are, to be sure, a natural event. But we have, from the beginning, been put into our own hands. One can see that this is so from the stunning remains of the art of the paleolithic period. We are, as Heidegger says, ontologically exceptional, the being for whom *being* is an issue. We must be concerned, first and foremost, with our own being and how we are living.

I share the broad environmental consensus that consumerism is, ecologically speaking, *the* problem. But why are we so committed to consumerism? Consumerism should be regarded as a religious attitude. It lives from religious zeal, a hunger for transcendence diverted into a dead-end. Consumerism is a perversion of Christian values. Once we start to realize that we are in the grip of some kind of pseudo-spirituality and pseudo-self-transcendence—people do not shop for things; they shop for a new self and a new future—we see that we are unfinished beings. We have a role to play in our self-completion. But we try to complete ourselves in an impossible way. We try to quiet the restless heart by shopping. We think endlessly about what kinds of products we are going to consume or what sort of experiences we're going to curate for ourselves. And no one is particularly satisfied with this—which is great for GDP, a system of endless economic demands and growth but an ecological disaster. It is secular Christianity that bears the burden of guilt here. We, as Christians, ought to know better. We are in a privileged position to understand what went wrong: Christian personalism became economic individualism in early modernity; Christian incarnation became materialism, Christian faith, hope, and love, became (in this order) social-economic narcissism, the cult of progress, and consumer envy.

MC: It strikes me that this worldview is, in many ways, eschatological. It synthesizes both the apocalyptic and the eschatological. There is, to keep it in the Christian, religious context, a sort-of original sin that humans bear: this ability that human beings have to be both the worst and the best. The very worst thing about us is potentially the very best thing about us. We have this capacity for greatness—at both ends of this spectrum. We can be profoundly destructive and profoundly creative. And there's that big breadth in the human being. I find this sort of strangely hopeful. We view the crisis as a crisis, to quote Nietzsche, in the way that a pregnancy can be a crisis. It's a sickness that is "pregnant with the future." I find that to be a very humanistic approach, one that provokes a kind of reimagining. A reimagining of what we are capable of and what we can be. And I wonder, psychologically, how this reimagining—our reimagining of the crisis this way—might change the way we understand the psychological toll that the ecological crisis has taken on us. There is this great psychological effect—people are of course being directly impacted by the ecological crisis. But what is the potential here, going forward?

SMG: Genuine eschatological hope must be distinguished from optimism. There might not be any hope for us, for late Holocenic consumers. Climate change activists do not like to say what they know to be true. We are destined to suffer the effects of a carbon legacy that will play out regardless of what we do now. Even in the most favorable circumstance of reducing emissions to net 0 by 2050, there is a climatological legacy that is going to play itself out in devastating ways. And the things that are happening in the world—floods, fires, and so on—are only going to increase. And yet we must also plan—not in spite of that imminent collapse but because of it—for a deep human future. As geology has no trouble imagining a deep future for the earth, we must become adept at imagining a human civilization 10,000 years from now, when the world becomes green again. There are many different ways to be human. And we seem to be stuck in a self-destructive pattern. We must free up the imaginative capacities in the human being in order to mobilize a will to change. We are being asked to make sacrifices for people who will no longer remember our names. And no other generation has been asked to do that. Aristotle can imagine an ethic that considers the good of the next generation. But beyond that, an intergenerational justice for distant future humans—this requires a whole new order of ethical commitment. This is a moral bind that we have never before been in. It touches the very core of our identity. I think the Gospel has something important to offer in this regard. Christians have been talking about the deep human future since the first century of the common era.

The eschatological attitude of the first Christians over-
took pagan Rome and introduced new values and attitudes
into the ancient world. The eschatological attitude of the
16th-century European reformers inaugurated modernity.
Imagine the political power of a widespread Christian eschatology with an ecological conscience! That is the not-
so-concealed theological sub-text of For a New Earth.

A Dialogue with Jim Morley

Matthew Clemente: I'd like to start off by hearing your thoughts on the theme of
"hosting earth." Of course, in many ways, The Guestbook
Project is interested in this idea of hosting the stranger and
being host means both being hospitable to the other—to
that which comes to us from outside—but also being willing to play the role of guests and the reciprocity that exists
between us and others and how we can reciprocate that role.
How is it possible to play host to the earth and be guests of
the earth and what does this mean for your research and
your interests?

Jim Morley: Well, you know it's a beautiful idea. The Guestbook Project is something I've been enamored of. It's been mostly
directed towards human conflict and this new initiative towards the ecological crisis is a needed frame. That we are
both guests of the earth and also engaged in caring for the
earth, as hosts of the earth, is a way of thinking that may
offer new insight. We need to overcome the linear dualistic
thinking that got us into our current crisis, and I find the
idea of "hosting" and "guesting" quite applicable because it
frames the problem in *relational* terms. And I think the only
way out of our current situation is to think in such terms. It's
the only language we can understand and it's the only frame
from which we can operate.

  Most of my peers in eco-psychology have talked about
the subject-object problem. A lot of them have been going
in the direction of changing our consumer behavior. Though
important, behavior change is, as we say in logic, necessary
but not sufficient. I'd like to go a little deeper with some
ideas that I was fortunate to be exposed to as an undergraduate when I was influenced by the social ecologist Murray
Bookchin. He opened me to ideas that have been germi-
nating across the arc of my life. The key leitmotif I'd like
to address, drawing from Bookchin's ideas, is the nearly

metaphysical, and even religiously held, presupposition of the inevitability of domination. We think of domination—and it's a term that gets thrown around a lot—in terms of a one-dimensional critique of the "bad guy" dominators who exist "out there." Instead, we need to go much deeper than that. Simplistically satisfying as they may be, moral pie-tisms only conceal, perpetrate, and exacerbate the problem. We need to, instead, exorcise this demon that lives within each of us, hijacks our minds, and runs our world, namely: *the presupposition of the necessity of domination over each other, and therefore nature.* My thesis is that we project our beliefs about our own human nature onto nature.

MC: Indeed, it does seem like there's this constant question about how sadism impacts our relations with one another. Are we going to try to possess the other? Are we going to try to sub-jugate the other? Are we going to try to force our wills upon the other? Or are we capable of seeing that our relation to others impacts our relation to the world itself? The earth both precedes us—we're born into it—and we also go out into the world and do things to it. Yes, it's very interesting how our dynamics—even within our family life—impact our relation to the world. We're born to parents and then we go on and have our own family and our own relations and sometimes we replicate in future relationships what's been either done to us or how we've related to others in previous relationships. And what's fascinating is that we don't do this just in relation to one another. But we do this in relation to everything, and in particular in this context, in relation to the world itself. I would love for you to unpack this a bit.

JM: A good example of this is our relationship with drugs. The War on Drugs has been an operatic dramatization of eve-rything wrong with our relationship with nature. We de-cided to *police* our own desires through the War on Drugs, and it has backfired monstrously. Consider the very idea of policing desires. All human history, since we ceased hunter-gathering and settled into hierarchically ordered urban social relations, is one ongoing panic over "*who is in charge?*" All of history is this perpetual panic. As a species we have evolved to this condition of consciousness—this amazing gift of awareness—and we've squandered it on the fear of *who's in charge*? There are many examples across the history of ideas that support this—the Hegelian Master–Slave dialectic is an archetypal example. If you think of it, the driving engine of human history is the terror of our freedom. We either rebel against someone controlling us or, in turn, we strive to seize control over others before they can control us. We have been living in this continual state of collective terror desperately grappling the serpent's head, but never able to let go for fear of being seized by that same serpent we are trying to control. We are condemned to a cycle of control. We play this out in our relations with our children and loved ones, but we also play this out in our relations with our own bodies, sexuality, emotions, illness, death, animals, and the natural environment generally. You see this emerging in Machiavelli. His social theory is essentially a manual for how to get and hold power. But you see it most clearly expressed in Hobbes, whose social pessimism became the worldview of most English-speaking peoples. He says very clearly that the social contract is based on a terror of one another. We have a social contract to protect ourselves from each other. A *policing* contract on each other. And the fundamental presupposition here is an almost theological belief in the inherent malevolence of humans. That is the core belief here: we believe that we are evil, and we project that evil onto nature. We believe that nature itself is something threatening that needs to be policed and controlled. *Bellum omnium contra omnes.* But there is an opposing view on the matter. William James, for instance, speaks of this in religious terms: "morbid-minded" spirituality versus "healthy-minded" spirituality in *The Varieties of Religious Experience.* He is thinking of course of the Quakers versus the Puritans in the American context. The Puritans have the Calvinistic notion that humanity is degenerate and inherently sinful and needing control, policing, and domination. This was the attitude that justified slavery and the brutally violent way the frontier was "settled." The Quakers have always held the opposite view. They believe that we are all God's children with God's light shining through us and, to the Quakers, we only need to wake up to this light that we always already are. This theological antagonism has played out across American history—and indeed across human history. When I teach Machiavelli to my students, they are attracted to his ideas—because they easily recognize his misanthropic ideas as their own. Machiavelli is intuitive to them. They recognize and enjoy having their own presuppositions reinforced by Machiavelli. Then, when I introduce more "anthro-philic" thinkers, I can sense their resistance to these ideas. They see positive attitudes towards human nature as overly idealistic and unpractical. So, we are up against a challenge here: the core belief that we're inherently evil. When Richard Kearney writes about Abraham welcoming the three strangers into his tent, we see here an act of trust and faith in human nature—in nature generally. It is an act that's basically an affirmation of decency. It is a risk, of course, but I think it comes down to this: the ethically foundational belief that I'm going to trust and have faith in the decency of strangers. This is an admittedly very tricky thing to do. But the next question is an applied matter of pedagogy: how do we cultivate and advance this faith in the decency of each other and, in turn and by necessary implication and extension—to nature itself?

MC: So, along these lines, what you've identified here is really this dividing line in how we view one another and ourselves and our relation to the world as a division between trust and fear. There is indeed a huge risk in trusting the other. We don't trust the other knowing that it's going to work out. We trust the other with the hope and possibility that it *might* work out, but we do this knowing the risk that it really might not. And in your comments, you note how we dominate and police one another, ourselves, and nature, how out of that lack of trust, that fear, we try to control and prevent things from happening. But as we see now in our own relationships—and in the ecological—the attempt to control often provokes the thing we fear to come to fruition. When we attempt to assert our will, we actually bring upon ourselves the crisis that we're trying to stave off. I'm curious about how it's possible to trust. We're talking about something that's hardwired into us—control and domination—and the world is much larger than we are. The world is not a brother but a stranger, to allude to Camus—the world is absurd and all of a sudden it reveals itself to us. And in that moment, we feel our own vulnerability. But how do we attempt *not* to control and dominate in that sort of instance?

JM: Camus, I think, provides the answer to this problem in *The Plague*. There he makes the case for human community as the antidote to our condition. We are living in a time, indeed, of plague—it is a crisis and a call to community, and indeed a way out. The crisis forces us into community, but, on the other hand, it can also force us into a police state. That's the nature of crisis—it could go either way. In terms of the trust issue: you are right, again, it could go either way. Trust is, in its very essence, a risk. And this risk cannot be based on blind trust alone. It needs to be based on something. You can help mitigate bad outcomes by encouraging trust in one another. But this call to trust can be supported through education in some basic scientific facts—facts that support the thesis that humans are intrinsically cooperative, empathic and inherently caring. There is biological evidence built into the morphological structure of the human body that speaks for itself. Evolutionary anthropologists have been pointing out for decades that there is evidence that—our very biology—is oriented towards affiliation and pro-social behavior. But getting people to believe these facts is difficult. Our taken for granted common sense "natural attitude" keeps us from even seeing the veracity here. But we need to break through these mistaken collective assumptions by doing a better job at presenting the evidence, whether that be through philosophy, art, literature, or music. Again, the evidence is clear: Darwin's idea of adaptation holds that we *adapt* to our ecosystem, and we have adapted to our ecosystem by practicing pro-social cooperative behaviors. There are many ways in which the human body is naturally structured to show how we have evolved as affiliative beings, and not aggressive beings. We have no offensive predatory affordances—no venom, snouts, horns, claws, or fangs. Our muscles are not particularly powerful, and we cannot run fast enough to ambush or escape most animals. In contrast, we are profoundly vulnerable. Due to our overlarge heads we are all born months prematurely and enter this world in a paralyzed condition of total dependency—manifesting in a life cycle of emotional neediness. Our unnaturally upright posture exposes our soft organs and puts stress on our knees, backs, and shoulders. Our globular heads are hard to balance, and we lack protective thick skin or even fur. This affiliative morphology, however, is pronounced: our genitals are rotated towards the front, which means that most sex is face-to-face, eye to eye, and personal. Our hands, lips, and erogenous zones are packed with tactile nerves. Our sexuality is ubiquitous: we are sexual before we are born and remain sexually capable to the moment of our death—and what is sexuality but affiliation? Most other animals (except for bonobo chimps—our nearest genetic relative) have fixed sexual response mechanisms that are only elicited when the females are in estrus. Again, in contrast, humans evolved to be perpetually open for sexuality, both emotionally and literally, in a way that is continuous throughout our lives. We perpetually parent our young and never detach, metaphorically, from the breast. At least symbolically, we are always breastfeeding our children. Unlike any other species, we care for our children beyond maturation, into adulthood, and even until the day we die. Perhaps the most profound aspect of human biology is what's called *paedogenesis*, or *neoteny*, which speaks to the fact that we are essentially *fetal* in our morphology. In other words, our morphology is fetal or juvenile in a way that permanently extends into adulthood. We retain fetal-like structures all the way into reproductive age and even into seniority. Put simply: we are perpetually youthful. No matter how old you are, your body retains youthful morphological features—as does your mind. The immense evolutionary payoff here is psychological. This somatic and emotional youthfulness addresses our so-called intelligence which I define as tolerance of ambiguity, curiosity, and playfulness. While other animals cease their playful open-ended attention structures once they reach maturation, humans never lose our capacity for open-ended play. We are, as Gadamer rightly says: "beings of play." The capacity for perpetual play and *learning* became our ultimate evolutionary adaptation that replaced any need for predatory or defensive somatic features. Humans are the beings who are always capable of learning. This cognitive and emotional flexibility defines us. Question asking, tolerance for ambiguity and capacity for—not just novelty—but *imagination itself*, are the evolutionary strategies for our stunningly successful adaptation to every environmental adversity as much as it also undergirds our extraordinary capacity for empathy, cooperation and mutuality with one another—and the natural world.

We are the most affiliative and caring species in the known universe. Kropotkin, a Russian anarchist theorist, was actually a biologist and evolutionary theorist as well. Kropotkin's field research in Siberia was fully at odds with the Social Darwinists of the English-speaking world who held that humans are essentially bestial, in the Hobbesian sense, and that aggression, competition, and domination is *natural* to us. Doing fieldwork in perhaps the most hostile environment on earth (Northern Siberia) Kropotkin found evidence that organisms naturally support each other in a mutually adaptive way. They develop *symbiotic* relationships with each other and the environment. When the environment is stressful, organisms support one another for mutual survival. He translated this evolutionary insight into an anarchist social theory that purports that, in our basic nature, we humans are supportive and find ourselves in mutual relation with one another in an intuitive way. Social cohesion is our evolutionary baseline—not an anomaly. I think it is vitally important that we promote Kropotkin's evidence-based findings instead of the ideological misinformation of "survival of the fittest"—a phrase Darwin himself profoundly disliked and expunged from his second edition of the *Origin of the Species* due to its appalling misuses by the "Social Darwinists" who controlled the social sciences of the early 20th century.

Another point here is that the most positive aspects of the world's religious systems are affirmed in this understanding of the paedomorphic or neotenous structure of our bodies. Religious mystics and spiritual masters have always intuitively understood this when they valorized the wisdom of the child-like point of view. The Machiavellian narrative that we need to control everybody and the Hobbesian narrative that life is nasty, brutish, and short is just not accurate. We need to break this dangerously incorrect collective misunderstanding about who we are. We need pedagogical strategies for breaking this "natural attitude" of our essential negativity and highlight the true narrative—how we are inherently affiliative and empathic.

MC: You provide a truly humanistic perspective—a humanistic ecology. In order to understand how we're to relate to nature, you've seemed to intimate that we need to understand ourselves. But if we are to understand ourselves, we must understand all aspects of the human condition. We must, at the very least, be interdisciplinary in our approach: we must understand—as you've mentioned—biology, religion, politics, psychology, art, literature, among so many other things. Help us trace a bit more the role that these other "unique disciplines" play in human life. How does understanding all these things help us understand human life? How do art and literature, for instance, help us better understand our relation to the world?

JM: Nietzsche famously asserted that: "Life is not worth living without music." Nietzsche is saying that, as the "sick animal," the way out of the pathology inherent to our human condition is through the community that comes with experiencing and identifying with the expressions of others and the integrity granted by creating art. As a psychologist myself, I have observed that all recovery involves expression—and art is, in its essence, expression. I fully affirm Merleau-Ponty's overall notion that "to exist is to express." And to be in psychopathology, i.e., to psychologically suffer, is to *not* be able to express to others and oneself, and vice versa, to be estranged from the intimate expressions of others. You could think of this almost in organic terms: organisms need to express. So, I think the way out, if you will, is relationship—connection. I do not want to melodramatize, but I think it fair to say that our hyper-aggressive and toxically competitive culture is what casts us into the psychic and ecological crisis of our current historical moment. Depression and anxiety rates are remarkable, especially among teenagers. The mental health crisis is real. So is the climate crisis. This is because we can only foresee spending our lives entombed into an office cubicle condemned to pointless jobs where there is little hope for meaningful self-expression, a meaningful future, control over one's life, but most importantly *connection* to nature and other people. It's an economy, technology, and culture that is structured to isolate and increasingly disconnect us from each other—and nature. It is a world designed for the production and movement of goods and services but not one designed for maximal human thriving which means extended families and face to face community and contact with the natural world. Ours is a world *designed against* the raising of children and the maintenance of extended family networks that are the baseline necessities for successful human development and thriving. It is through the security of kinship networks that we learn to express, and receive, the intimate expressions of others. In an aggressive, impersonal, competition-based culture like our own, it is interesting to watch the eradication of art from the curriculum of so many schools. Art is essential to human thriving and our ability to function as expressive human beings. I am not only speaking of "high art" on the level of say, Picasso. I especially mean everyday handiwork or *hands-on* skills and crafts. It can mean finding pleasure in any form of handiwork—from woodworking, to mechanical repair, plumbing, or cooking. Even advanced engineering or technology such as computer design can involve purposeful decision making and meaningful expression. To create is to be connected to creation.

One of the biggest threats to our ability to solve our current crisis is *insularity*, especially in the ivory tower of the academy. We academics are the gatekeepers to knowledge, yet we do not always have the courage to pass through those gates ourselves. Worse, we too often block others from passing though those gates. We know that the disciplinary turf wars we propagate in academia are profoundly wrong—but we do not know how to stop ourselves. Just like our American Congress, who all know exactly what is wrong with what they are doing, but don't know how to stop the monster they keep feeding. As demanded by the student protests of the sixties, we need an interdisciplinary academic model that values the polymath over the specialist and the independent thinker over the obsequious one. We have models in many continental thinkers, such as Merleau-Ponty, who applies the physical sciences to illuminate phenomenological ideas and vice versa. But art for Merleau-Ponty is the primary paradigm of all human reality. It is the ultimate paradigm. This is key. Yet, unfortunately, in our utilitarian society, the physical sciences, in a short-term way, deliver what our current vision of economics holds as most valuable: military and economic power and domination. We need to be aware of this if we are to design strategies for moving on.

The emblem of contemporary humanity is the *astronaut floating in space*. Encased in a sealed suit, touching nothing, suspended in the empty void, hovering outside the earth, we moderns find ourselves drifting over an abyss detached from life itself. Educators and health care professionals understand that the real pandemic is not viral—it is psychiatric. We are not depressed because of something wrong with our brains, but because we lack relationship, rapport, affinity, empathy, or connection to others and the natural world. In a word: loneliness. In our industrialized auto-mobile suburbanized social vacuum, we are all floating astronauts untethered from connection to history or any "place" that is a home.

Facing the approaching cataclysms of nuclear proliferation, the consumer waste economy, climate change, agricultural/industrial toxicity and social inequality, it becomes all too easy to default into pessimism. But I would appeal to the spirit of Camus's Sisyphus, who, condemned to a hopeless task with no end in sight, still chooses to roll that rock up that hill in defiance of the impossibility of his situation. He commits himself to rolling that rock. So, like a patient with a life-threatening disease (a potent metaphor for our collective predicament), it is not the disease itself from which the prognosis is made, but the patient's *attitude* towards the disease that counts most in the recovery process. Like that patient, we have roughly three options: we can ignore the catastrophe around us, we can surrender to the hopeless inanity of the situation, or we can engage, like Sisyphus, and take up the struggle—regardless of the odds.

So, while I obviously endorse the path of commitment, any struggle still needs an accurate assessment and an effective strategy for action. What I am offering here is my diagnosis—which is that the ecological crisis boils down to how we see nature as separate from ourselves. We see nature as other—as an object, a thing, a stranger. This is the first level of diagnosis. The next level is that we see nature, this ultimate alterity, as a hostile and malevolent threat. Thirdly, and most intimately, is the core issue of how *we reciprocally assume ourselves to be malevolent and dangerous*. This is the vicious psychopathological circle that I think needs to be addressed before we can take any first step towards action.

It may be the essential tragedy of our human condition to be estranged. Whatever our original evolutionary condition may have been, human consciousness, and our subsequent process of civilization, invoked a radical separation from nature that has been haunting us from the beginning. In other words, with the evolution of our miraculous gift of consciousness came a shadowy dark side which is that we see ourselves as separate and thus alienated from our environment. Taking up the metaphor from Genesis, when we ate from that fruit of knowledge, and woke up into the godlike gift of consciousness, we immediately felt ourselves severed, distinct entities, no longer immersed in the participatory totality of nature. As individuated beings, now aware of our personal mortality and duly terrified by its implications, we perpetually perform this traumatic split between ourselves and the natural world, a split which Descartes and most modern philosophical systems have only been the messengers.

Our pre-modern ancestors knew something we have forgotten. They addressed the problem of our essential pathological condition of existential isolation by developing soothing religious mythologies, rituals, and community practices to help us to remember who we are and from whence we came. Shamanic spiritualities maintained an imaginary enchanted relation to nature, remedying our sundered consciousness through healing arts—a relation that modern abstract theologies no longer address. Our shamanic tribal ancestors, contemplative monastics, and mystical practitioners knew how to enter the visionary dreamtime that reunited us with the perpetual present moment, or now-point, of nature. Our ancestors knew how to address our pathologically alienated condition with what is called "medicine" by indigenous peoples. This was a psychological wisdom that is lost to modernity and only partly retrieved by a psychiatry still inhibited alas by its Cartesian straitjacket.

We must be careful, of course, to avoid the awful mistake made by fascists which is to assume an unreflectively romantic nostalgia for an idealized past. We do not want to go back to, for example, human sacrifice and ritual violence, to sentimental naturalism or reactionary biologism. Nor do we want to lose ourselves into a regressive collective identity. Instead, we have the privilege to be selective about the technology and social practices best suited to our real social and ecological needs. We can maintain the benefits of the enlightenment ideal of a universal humanity—as much as we can select the technology that is maximally 1. socially liberating and, at the same time, 2. ecologically sound. In this social-ecological approach, human freedom, technology, and ecology are mutually interwoven. Here, my freedom is contingent on the freedom of others as much as my existence is contingent on the conditions of the ecosystem. Again, industrialization and political freedom need not, in any way, be in an aversive relationship. The difference is mindset and point of view based on the assimilation of evidence-based science and a clear articulation of what is truly valuable. If our goals and purposes are set right, we will see the world in a radically different way. Like Husserl's famous attitudinal shift that is made possible by the methodological *epoch*. We could name and suspend our utilitarian and ecologically destructive presuppositions to clear the way for a social-ecological perspective. Put another way, we could call this a social-ecological system of hermeneutics where we actively apply our ecological values to our choices in technology and social practices. A social-ecological approach would creatively appropriate the technology we need from both the past and the present for another kind of future world—which is possible.

MC: Any concluding thoughts?

JM: Environmental, social, and psychological change are inseparable. The evidence contradicts the commonly held belief that we humans are inherently aggressive. Like the yogic notion of *karma*, our deeply held beliefs, like self-fulfilling prophesies, become manifest in the actual world. Because these negative social beliefs are projected onto the natural world with pathological consequences, it is a precondition of any environmental policy that this dangerous misinformation about our own human nature be confronted, refuted, and removed from our worldview. This false narrative needs to be replaced with an interdisciplinary and integrative social-ecological worldview based on the truth that we are affiliative and inherently empathic beings. Addressing our self-understanding, in this way, will reform our attitudes towards nature and progressive ecologically sound behavior, technology, and economic activity will follow from this new attitudinal shift. We cannot do unto nature what we cannot do unto ourselves. We cannot relate to nature in a positive way if we cannot relate to ourselves in a positive way—and vice versa.

A Dialogue with Ed Casey

Matthew Clemente: I'd like to start by asking you to talk about how you understand the theme of *hosting earth*: what does it mean to play both host to, and to be hosted by, the earth? How do we situate ourselves in that relationship?

Ed Casey: I'd like to speak about how I came to be particularly concerned with the natural world and its phenomenological significance. I come from Kansas, a world of farmlands, flint hills, and feeling very close to nature, since the time I was young. However, as I moved along in my educational career I retreated into libraries, study halls, carousels, and as a young philosopher, I felt my task was to write. And write I did. But that really meant leaving behind the natural world in any significant ongoing intense way. I was in a period of retreat for about twenty years. From my high school years, when I first discovered philosophy, until my early thirties. And then my writing took a sudden turn that brought me back to the earth, if you will. I was writing a book on memory and I wanted to write a chapter called "Memory of Place." And the more I thought about it, the more I realized that in Western thought—after Aristotle—there's a paucity of serious thought about place, particularly place *of* nature and place *in* nature. That realization changed everything. I put down all my other plans for writing and began to write about place. And place took me, of course, straight into the natural world. My first book on place—*Getting Back Into Place*—appeared in 1993 and has a section entitled "Wild Places." My focus in the book was on "moments of nature," moments that included such terms as "sensuous surface," "ground things," "arch-atmosphere" and so on. I was engaging in phenomenological description, but instead of seeking essences, I was seeking descriptive themes that populate the lived experience of nature. It was a matter of getting clear about how the natural world presents itself—and in what thematic formats—as a place. A place that is different, especially, from urban settings.

After writing this book, I became a charter member of the Environmental Philosophy Association (closely associated with the Society for Phenomenology and Existential Philosophy) in order to examine the plight of the earth in more graphic detail. But instead of turning to environmental activism, as was often the focus of early meetings of this group, I continued to explore the natural world through a series of books whose focus continued to be place (in terms of representing place, landscape painting and maps, and whatnot). My work took on the question of what happens to the earth when it assumes the format of images of landscape, painted or mapped. I know from my own painting; landscape is both affirmed as presented *and* transformed: both at once. I think about how artists reshape landscape, extending from Robert Smithson to Willem de Kooning.

In my more recent work, I consider how edges configure the earth as we encounter it directly with our bodies. Edges of rivers, mountains, lakes, ponds, and parks impact us so heavily. I have come to realize that the surface of the earth cannot be what it is except insofar as it is shaped naturally or, indeed, artificially. The earth can be forced into shapes that it does not wish to bear. I am interested in place—the spatiality of place—and its multiple avatars. Looking back, I can say that philosophy initially took me away from direct experiences of the earth (which I knew when I was very young in Kansas) but eventually brought me back to the earth, especially through my pursuit of phenomenology of place. "Back to the things themselves"—or, rather, back to that which underlies and precedes all finite natural things—the earth.

There is an affinity between phenomenology (especially in its post-Husserlian modalities) and the dyad of nature and place. In many respects, this is the converse of Jacques Derrida's radical departure from Husserl. A departure that led us to favor such things as "grammatology" and "trace" and "text" all in the context of aporetic indirection, rather than direct description. But if Derridean deconstruction plays at the convoluted surface of discourse, a phenomenology of nature dives down into the earth. We can say that it hosts the earth from the word go. In this spirit, we can say that writers and painters have been hosting the earth through their own work for millennia.

MC: You consider how, initially, writing took you away from place, from the lived world of nature which you grew up in. As you were saying that I thought of the analogy Plato makes at the end of the *Phaedrus*: writing as analogous to farming and to tilling the earth and sowing seeds in the earth, wherein writing can be this kind of thing that bears fruit. He uses some really fruitful nature imageries to describe the process of writing. And yet, at the same time, writing can similarly build walls and barriers. I think of art—writing and painting—in relation to nature. It can open us up to nature and also potentially move us away from nature. For example, I was just reading Unamuno's novel *Fog*, in which one of his characters is sitting in a park in the middle of the city, looking up at the trees, and the narrator reflects, "Those domesticated urban trees, standing at attention in straight lines, watered by an irrigation ditch when it didn't rain, extending their roots under the plaza's flagstones... Those imprisoned trees waiting to see the sun rise and set over the rooftops... Those caged trees that perhaps yearned for a distant forest... Augusto felt mysteriously drawn to them. Birds sang in their branches, city birds that learned to flee from children and sometimes approach the old folks who offer them breadcrumbs." So there is this notion of trees that are something that exists both in nature and for the world and also something that we domesticate, which is not something that we typically think of. We bring them into human constructs and spaces as we build around them and cultivate them. And I know that this has a kind of resonance with some recent work that you've been doing.

EC: I am writing a book called *The Place of Trees*. One of the striking things that has emerged in my writing this book is the way in which trees communicate among themselves and have a social life that has only been unearthed in the last decade or so. It is in that respect that trees have highly sophisticated and very semiotic ways in which to talk to each other. I contemplate how this modifies our idea of place as merely *stabilitas loci*; that is, providing only stability and anchorage. But the place of trees is very highly articulated and animated even if invisible and underground. We have not fully understood this life—our exploration of it has only just begun.

But let me return to your earlier remark about the *Phaedrus*. I have much sympathy with that—in fact, it was when I was in my mid-thirties that I first realized that my writing wasn't getting all the way "there." And at that moment in my life, I turned to painting. I discovered, at that point, that there was another way into the natural world other than language and verbal discourse. I took up painting in a serious way and rented a house in Maine one summer—along with a group of fellow painters—and began to paint the earth. It was an inroad into understanding and interacting with place—but it was nonverbal. For me, painting opened up a second path to the natural world. And the natural world seemed to be inviting my efforts. Painting, unlike writing, had a sort of freedom for me that writing—at least prose writing—just didn't have. So this has come as a sort of thematic in my approach to nature. I paint just as much as I write. And this, or some comparable direct engagement with earth, is something that philosophers of nature—phenomenologists of nature especially—must take seriously as they consider the ways in which we interact with, and reflect upon, nature.

MC: A theme that seems to resonate here is that one of the ways in which we can be more hospitable to the earth—and feel hosted by the earth—is to engage in artistic forms of creation (forms of creation that are not exploitative, but of course that attempt to honor the world as it presents itself

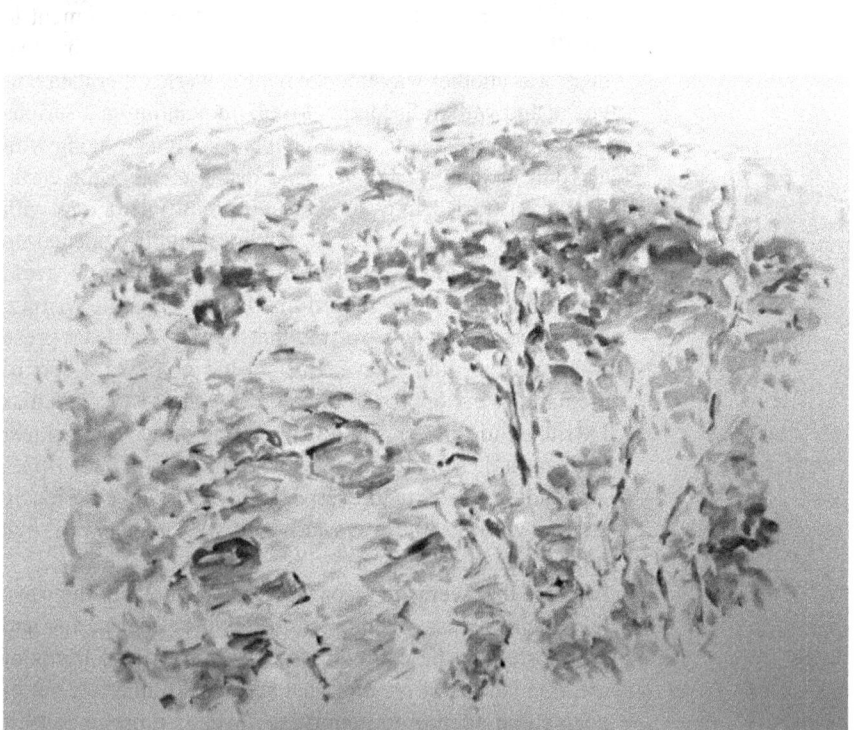

to us). One topic that seems related is how art functions in understanding place. What is it about art—the creative process—that helps us understand our situated-ness within the world and to appreciate that fact in a way that's maybe different from the temptation to exploit the world. What makes art and poesis different?

EC: The way in which the lived body is solicited to create free motions that are not confined to the linear as is writing can liberate one's way of ingressing into nature in a far more direct way than is possible with writing. With prose—in Merlaeu-Ponty, for instance—there is certainly a beautiful way in which nature can be described. But still, it is not the same as the gestures—the motions and actions—that the lived body makes when it is painting or sculpting. So I consider artistic activity not merely supplementary but also an essential complement to writing about nature. Philosophers, including philosophers of place, should indeed explore these alternative methods of moving into the natural world and opening it up from within by dint of their own bodily gestures in a way that cannot happen sitting at a desk in a cubicle confined by the page, line, syntax, and the massive overdetermination at play in writing philosophical prose.

MC: So then it seems as though by returning ourselves to bodily existence, we can come back to ourselves as bodily, carnal creatures. Through that movement we move back towards nature, which of course is physical space and proximity and whatnot. And so, it's by recapturing carnal existence that we move toward a different type of way of relating to nature than we're potentially used to.

What Richard Kearney calls "touch" in his book of that name is certainly relevant here. It's not only seeing nature or thinking about nature, but also the tangibility of the art-work that becomes the felt analogue of immersion of the lived body in the natural world itself. And this is not just a matter of sublimation, it is really a matter of direct experience. There is an ongoing "living experience" in painting the natural world. I paint seascapes, landscapes, mountain-scapes. I always start from the natural world, but I also always depart from that and go towards something that can be considered abstract. As if to encompass these two aspects of my own life—the embodied and the abstract.

Chapter 6

Climate Emergency and Radical Ethics: Colonialism, Racial Injustice, and Climate Justice

Donna M. Orange

The scientific information about climate now describes not change but emergency. The political will to face this reality has diminished. Though it sometimes seems resurgent, it fails to stick for reasons we can all see. My climate psychology alliance friends are hearing voices like those of Jem Bendell who writes of deep and humane adaptation as we prepare for the imminent end of a habitable planet. David Wallace-Wells is writing pieces like "time to panic." The big question has now become: do we have time to change course, even if we can summon the political will? If not, how should we live for the time being?

So, let us begin as if we still had time.

Psychoanalysts, though more alert now to our responsibility to the world's most vulnerable people, more conscious of our solidarity with those who suffer, still seem to be working largely in a bubble. Climate change has already, scientists tell us in the most urgent voices they can find, become an emergency, threatening to overwhelm all attempts to stem the primarily human-created disaster. Still we psychoanalysts work quietly and faithfully on, living as we always did, driving and flying less but primarily due to COVID, watering our lawns, eating and consuming mindlessly. Meanwhile, most political and financial leaders conspire to hide ominous truths, no longer simply inconvenient but dire, and we allow ourselves not to notice. Are we psychoanalysts who should perhaps do better, conspiring or colluding to sustain an environmental unconscious? Are we helping to silence the canary in the coal mine?

Having received extensive education and training, including mandatory personal analysis, to prepare us for our work, we have, I believe, also acquired responsibility to be leaders, moral if not scientific, in confronting the global crisis we are living. We possess the intellectual and communal resources to take on this responsibility. So far, however, we have been resoundingly silent.

Where are the psychoanalysts, we who, rightly or wrongly, consider ourselves intellectual leaders in psychotherapy and in understanding human motivation? Perhaps we have learned nothing from the example of Sigmund Freud, who, blinded by his passion for his work, his love for Enlightenment German culture, and his need to be as important as Copernicus and Darwin, could not see that he and his Jewish family, as well as psychoanalysis itself, faced mortal danger in Vienna

DOI: 10.4324/9781003456940-9

in the late 1930s. In another strange example, a few years later in the London *Blitzkrieg*, during one of the British Psychoanalytic Society's furious disputes about the origins of hatred and aggression, Donald Winnicott noted their actual effects: "I should like to point out that there is an air raid going on" (Grosskkurth, 1986, p. 321). Are we, too, so absorbed in our theories, and worse, in our theoretical and interdenominational disputes over who belongs and who does not, that we fail to notice that human-caused planetary warming threatens to destroy the world within which we practice our beloved profession? We say that all is grist for the psychoanalytic mill, but what if this crisis threatens the survival of the mill itself?

We psychoanalysts, I believe, together with our colleagues in other therapeutic areas, actually have a unique contribution to make in this crucial moment. We can help not only to refocus our own attention on the imminent threats to our own way of life, but to the world's most vulnerable people and to the earth which supports us all. In the best psychoanalytic tradition we can notice the forms of historical unconsciousness, the still-walking ghosts of the narrative unconscious (Freeman, 2012), keeping us insensitive to the suffering in which we are implicated and for which we are responsible. We can call out the more selfish of the defenses that keep us avoidant and name the forms of traumatic shock that keep us too paralyzed to respond appropriately. We can help with the processes of mourning not only the remembered ways of life, but also the loss of many kinds of hope and certainty for the future. Learning from Latin American climate justice leaders, from First Nations leaders in Canada and elsewhere, we can ask ourselves and each other— including our patients—what really matters in time of crisis, thus responding more creatively than our analytic forebears did. But we have no time to waste. In Bill McKibben's (2014, p. 41), words, "there no longer is any long haul."

I will skip most of the science section for two reasons: 1) I expect that this group already knows about the rising sea levels, the disappearing ice, the fast-increasing desertification, the intensifying violent storms and the fires, and 2) it gets exponentially worse every month, with the news outstripping the predictions. But for many years now even the NOAA and IPCC scientists have emphasized that:

> Risks are unevenly distributed and are generally greater for disadvantaged people and communities in countries at all levels of development.

Significant climate changes already underway will require major adaptations, especially to protect the most vulnerable people. Whole island populations must relocate in the face of rising sea levels, while refugees from war and famine are already pouring into Europe (Ryde, 2010). Without radical changes away from fossil fuels, including substantial decreases in consumption, initiated immediately, warming of 4 degrees is both extremely probable, they emphasize, and will make our planet uninhabitable by the end of this century. Like the IEA (International Energy Agency), the IPCC warns that every level (world, regional, national, sub-national, local) must make radical changes immediately to avoid the direst consequences. This prediction from 2014 is now considered seriously outdated. The weakness of international climate accords and national regulations means that the whole world is on a fast path to destruction.

Most of us go on just as if we had not heard these warnings. Are our brains simply wired to exclude bad news, as George Marshall (2014) believes, are we in the grip of "environmental melancholia" as social scientist reader of psychoanalysis Renee Lertzman (2015) thinks, or frozen by climate trauma as I have wondered (Orange, 2016)? Or have we inherited a philosophical egoism, wedded to a narcissistic mindset full of entitlement, perhaps even of unconscious racist privilege inherited from millennia of slavery, trapping and immobilizing us? All may be true, but if the disciplines of philosophy, history, and psychoanalysis can help us to identify our problem quickly, then perhaps alliance with the world's moral and religious leaders can begin to shift the political tide, to create a "tipping point" of solidarity to meet the carbon tipping points already looming. As you will hear from Jeremy, the determined courage of the world's young people may be now our best possibility. We need to hear them.

Enlightenment Egoism

We Westerners inherit an outsized share of the guilt and responsibility for climate change. Whatever we may think of China now, we in the West set the industrialization-at-all-costs pattern they have followed. As Stephen M. Gardiner writes, "the USA is responsible for 29 percent of global emissions since the onset of the industrial revolutions (from 1850 to 2003), and the nations of the EU 26 percent; by contrast China and India are responsible for 8 percent and 2 percent respectively" (2011, p. 315). To understand what has gone so wrong in our relation to the earth, including our indifference to its most destitute people, we must mention the egoistic roots of the scientific rationalism and political individualism emergent in seventeenth and eighteenth century Europe. These became founding ideals in the United States.

We must also remember that the climate crisis affects us humans unequally, that some of us are still reasonably comfortable while already many, and increasingly a large majority, remain hungry, homeless, and destitute. Some of us have far more coping resources than do our brothers and sisters across the tracks or across the world. Some of us can move away from rising water, from famine, and from war. We might say with Emmanuel Levinas, phenomenologist and prophet whose radical ethics we take up later, that to respond ethically, we must allow ourselves to be traumatized, taken hostage by the useless suffering of these others. From the Talmud he quoted: "To leave men without food is a fault that no circumstance attenuates; the distinction between the voluntary and the involuntary does not apply here" (Levinas, 1981, p. 201). In other words, not knowing that my lifestyle creates starvation and misery for others is no excuse; the fault remains. (Thus, climate justice theorists speak of "carbon debt" that polluters like us owe to those whose health and survival continues to be destroyed by our way of life, whether we know it or not.) Levinas commented: "Before the hunger of men, hunger is measured only 'objectively'; it is irrecusable" (1969, 201). He often quoted from Dosto-evsky's *Brothers Karamazov*, "We are guilty/responsible of all and before all, and I more than all the others." Double-mindedness, even when traumatically induced, does not exonerate. It simply leads to "we didn't know anything" from people who saw their neighbors loaded into boxcars and taken away. We too are pretending to ourselves not to know, especially how our lives of privilege impoverish and endanger our fellow human beings. Do we realize that many, if not most, of the refugees that Europe and the US are turning away are climate refugees, and that there will be many more, even here?

But traumatically paralyzed, we may not notice our guilt and responsibility. Or we may feel so overwhelmed by the outsized proportions of this crisis that we cannot imagine where to begin, and find ourselves just going on as before. "The fierce urgency of now" (King, 1967) worries many of us: by the time we wake up, millions more of the world's poorest will have died from hunger, and global warming's worst effects may be irremediable.

Historical Unconsciousness and the Invisible Present: Settler Colonialism and Slavery

Unconscious and silent about the US history of settler colonialism, ignorant and mute about our crimes of chattel slavery and racial domination, neither governments nor citizens can seriously tackle climate injustice until we confront this 400-year history.[1] This is my thesis. It accuses the "me," ever guilty and responsible, but locates the problems in a shared historical and narrative unconsciousness, in a "we." So let us imagine another crowd of ghosts inhabiting our ethical unconsciousness. Though intellectual ghosts may live to cause trouble—think of Descartes and Locke—now let us consider the historical unburied crimes, our collective ghosts, though largely unconscious, of settler colonialism and chattel slavery in the Americas, and especially in the United States, founded on the explicit ideal of human equality. I choose colonialism and slavery because our habits of keeping these crimes—genocides before the term came to usage after World War II—hidden from ourselves, may also be keeping us unaware of the impacts of climate change on people who live out of our daily sight.

The history in North America, where we meet today, begins with two nations, England and France, competing to rule a land allegedly empty of human beings, according to the doctrine of *Terra Nullius*, first invoked by the Spanish and Portuguese. Then came the young United States with its "Manifest Destiny" and by an expansive Canada, destroying the indigenous peoples. Canada, it seems, has recently attempted much greater acknowledgment of its "First Nations" than we have done in the United States, where we name them "Native Americans," oblivious to the preference of most for being called by their tribal names, or simply "Indians."[2] Settler colonialism does not simply manage colonies as Britain did in India, arrogant as this process may be, but replaces the peoples living there, as in Australia and New Zealand, Canada, and what became the United States, as if the beings living there were not fellow human beings whose lives mattered at all. Judith Butler (2010) would say these lives had been treated as "ungrievable." Calling ourselves "a nation of immigrants," we do not even notice their loss.

Chattel slavery names the system, including in particular the African slave trade of the fifteenth through the nineteenth centuries, of claiming to own, buy and sell, inherit, and force to work—usually by extreme violence—human beings. Other forms of slavery, including those practiced by the Nazis and in the Soviet gulags, as well as forms of child labor, sex slavery—unspeakably reprehensible as they are and the cause of untold human suffering—are not under direct discussion here. My argument is that chattel slavery, the "peculiar institution" that formed the United States Constitution, giving outsized power to the slaveholding South, and haunting us to this day (Davis, 2006; Blackmon, 2009), has dulled our moral conscious-ness to the point of unconsciousness. What we do not know, really profoundly and extensively and personally know, Freud taught us, we are bound to repeat. Together with the colonialist past we all share, this history of slavery and its ongoing effects, of which we rarely speak, blinds us to the misery that our carbon-and-methane spewing lifestyles are creating in the global South. We are repeating.

For many of us, this history briefly became real in the film, *Twelve Years a Slave*, making us realize that massive human rights violations have not occurred only elsewhere in Germany, Austria, Chile, and South Africa, for example. We in the United States live every day from the history and continuance of precarity (Butler, 2010) generated by the conscious and unconscious racism with which Ta-Nehisi Coates now confronts us. Writing to his son, he reminds us:

> enslavement is not a parable. It is damnation. It is the never-ending night. And the length of that night is most of our history. Never forget that we were enslaved in this country longer than we have been free. Never forget that for 250 years black people were born into chain—whole generations followed by more gen-erations who knew nothing but chains.
>
> (2015, p. 70)

Actually we might consider whether Europe and North America, especially we in the United States, suffer from a superiority complex. Symptoms of such a "com-plex" might include 1) the presumption that we own land stolen from indigenous peoples who lived upon it communally just because we bought it from people who "owned" it before us; 2) the presumption that the ways we who think ourselves white do things are the right ways; 3) the presumption that others worldwide should learn English, while we have no obligation to learn Spanish or other widely spoken languages; 4) a general insensitivity to our own arrogance and sense of superiority, easily degenerating into violence against those we consider inferior; 5) an incapac-ity to imagine ourselves into the predicament of those to whom we feel superior, thus a blunting of empathy and compassion; 6) a sense that the earth belongs to us, the so-called whites, and that others, "they" exist to serve our economic interests: to mine the minerals we want or need, to make us cheap clothes, to work at below poverty wages, and so on. This complex, with its embedded assumptions, largely unconscious and invisible, forms a web of life, generating comfort among those who carry it and creating death and fury among those we dominate. We do not know that we suffer from it in our fundamental humanity.

Psychoanalysis and Double-Mindedness

By double-mindedness, however, I mean something much more problematic than mere denial. Double-minded, we live in two realities at once. We know about the climate crisis, and perhaps even that it is rapidly becoming dire. But in a mental gesture akin to throwing up our hands, we say, this is too big for me or for any individual. Only systemic changes matter. So we go on working with our patients and our psychoanalytic politics, as humanistically as we know how to work, while our common home becomes a burning world (Cushman, 2007). Essentially, we, like Freud in the 1930s, and like the millions of "ordinary Germans" in the period studied by Thomas Kohut (2012) and others, have disabled the fire alarm. To accomplish this, as we shall see, we have taught ourselves not to see those worst affected as human just like ourselves, as others whose suffering matters.

We could spend chapters, articles, and books asking how this happens. A vast literature on multiple types of dissociation (Bromberg, 2006; Chefetz, 2000; Herman, 2009, 2011; Howell and Itzkowitz, 2016) and Freudian repression (Brenner, 1962; Jacobson, 1957; Loewald, 1955; Volkan, 1994) is now available to us. More immediately useful might be to ask how, faced with crisis, we, living daily in our professional minds with these concepts, might emerge rapidly to protect our common home. Will our theories help us enough—surely they have not until now—or do we need something else to shake us out of our analytic slumbers? Speaking to psychoanalysts, who often characterize ourselves not as belonging to the STEM (science, technology, engineering, mathematics) disciplines but to the tellers of stories, philosopher Judith Butler (2005) writes:

> But what if the narrative reconstruction of a life *cannot* be the goal of psychoanalysis, and that the reason for this has to do with the very formation of the subject? If the other is always there, from the start, in the place where the ego will be, then a life is constituted through a fundamental interruption, is even *interrupted prior to the possibility of any continuity.*
>
> (p. 52)[3]

My guess is that we cannot see, cannot feel this prior interruption of the other, primarily because we are so embedded in the cultures that have created the climate problem. As many have noted, Freud grew up intellectually in, and was deeply attached to, the Enlightenment and Romantic European culture that enshrined individual autonomy as the ultimate good. We psychoanalysts, and most psychotherapists, have inherited his deep assumptions, along with a certain blindness to threats to the "common good." Additionally, attached to these old cultures in part because of the traumatic background of contemporary psychoanalysis (Kohut, 2003; Kuriloff, 2014) creating a desperate need for home, we may have forgotten the other destitute homeless of the world, in particular of the Southern hemisphere. We need a contextualist analysis of our double-mindedness, and a radical ethic to shake us out of it.

Climate shame, one form of this double-mindedness, takes, I believe, two principal forms, intricately interrelated: 1) the fear of visible vulnerability that keeps us indifferent to the nature and extent of the crisis; and 2) Continuing envy of those who have more, so that those deprived of essential needs become invisible. In our striving we reach for more space, larger houses and cars, more things, perfect bodies that resist growing or appearing old. Entangled together, these two aspects of shame feed self-enclosed preoccupation, utterly distracting us from the descriptions and warnings of the best climate science. The first, the fear of vulnerability, keeps us self-protective and enclosed, endangering us humans, other species, and our planetary home. Without the fundamental experience of being held and cherished, we lack the emotional sense of belonging among other humans, or anywhere. Thus we treat our world, and implicitly ourselves, like disposable garbage. The second, envy, attempts to restore an ersatz well-being, to cover up the rotten sense of shame with fame, money, and glamour. Any sense that others have more occludes our ethical vision, blinding us to our involvement in injustice and deafening us to the cries of those we are harming every day. Much less can we, envious of those who have more, notice that we in the "first world" live as beneficiaries of massive injustices perpetrated by our colonialist ancestors on the indigenous "first peoples" in the Americas, Australia, New Zealand, and elsewhere, as well as on prosperity originally created by systems of human bondage also called slavery. For this unconsciousness to fade, or to wake up morally, we would need that indispensable and painful moral form of shame, awareness of our implication in systems of evil, to replace the shame and envy that traps us in compulsive consuming.

Just as psychoanalytic treatment of shame begins to restore to the mistreated a sense of inclusion in the human community, so can a psychoanalytic sensibility, sensitive to the corrosive and isolating effects of shame, begin to link us all with each other. We can begin to understand that our well-being depends on the well-being of others (solidarity, *Ubuntu)*, and restore the old notion of the common good.

Radical Ethics and Politics

Inverting all standard ethical accounts, radical ethics upends moral reasoning by starting from the suffering Other invading my self-satisfied life (M. Smith, 2011), not from the putative moral agent. Emptied of ego by a pre-originary responsibility, I am commanded by the other's starvation. William Edelglass (Edelglass, 2012) provides a compact argument for what I am calling a radical ethics of climate change:

> Because Levinasian ethics begins in, and ever returns to, the suffering other, the suffering of those whose lives are negatively impacted by climate change

rupture arguments that justify, or justify neglecting, individual GHG [greenhouse gas] emissions. Levinas acknowledges that in nourishing ourselves we take food that can feed others.

(p. 210)

In other words, my irrecusable responsibility[4] to the vulnerable and suffering other does not go away just because I need to drive or fly somewhere. Nor is it lessened because this vulnerable other ekes out a bare existence in the southern hemisphere, less visible to me. Philosopher Robert Bernasconi (2010) writes:

A hungry child in a distant land is no longer out of our reach. If globalization means living in a world in which the terms far and near, stranger and neighbor, no longer have the same meaning for us as they once did because everyone is now recognized as within the same circle, then the hunger of the most marginalized members of the global society must become its fundamental reference point.

(p. 77)

The other's precarious life (Butler, 2004) questions me, accuses me, persecutes me. A close consequence of this endless responsibility links it to the political.

"There is no place," writes Bernasconi, "for an ethical discourse that is not also inextricably linked with a recognition of the political context that it is its task to interrupt" (Bernasconi, 2006, p. 256). He explains that experiences of persecution link us to others enslaved, impoverished, homeless, in solidarity. Watching the thousands of families seeking shelter in Europe in 2015, those who remember their own suffering may be better able to respond. Instead, oblivious to the causes of the misery that brings children and adults to our southern border here in the US, we compete to boast about who can build the largest and most robust wall against them. Most of us, unless affected by tornadoes, hurricanes, and wildfires, have no memory of this kind of suffering. When the sufferers look much like us—perhaps white and even affluent, we rush to help them rebuild. But many of those devastated by hurricanes like Katrina are already poor and/or dark-skinned, so we treat them just as we treat the refugees. How do we come to see them as our brothers and sisters? Can memory, either of our own sufferings or of past courage and generosity, begin to activate us in the nick of time? In Merold Westphal's words, "Conscience need not make cowards of us all; it can make us brothers and sisters" (Westphal, 2008, p. 133).

This radical ethics of response to the other implies disrupting massive political injustice[5] when we are able to see it. It means that others' suffering persecutes me, takes me hostage, requires substitution, one-for-the-other. It demands hospitality to people displaced by war, poverty, terror. Even if we cannot speak their languages, we can see their desperate faces. We must remember. In our own history, we can remember quiet activists sitting at segregated lunch counters, sitting at the front of segregated buses, disrupting injustice. Now we find indigenous peoples of Canada, the USA, and elsewhere, sitting down to close the roads to tar-sands drilling equipment, standing in solidarity against injustice. How do we who live more comfortably find our ethical bones? Even when the photo of the drowned three-year-old refugee on a Mediterranean beach stops us for a moment in our tracks, how do we keep our ethical imagination awake, determined that a world in which some lives are regarded as expendable cannot continue? We need an ethic that arises from extreme situations, because we face one now. Let us hear these words of a survivor of five years in Nazi camps:

> justice remains justice only, in a society where there is no distinction between those close and those far off, but in which there also remains the impossibility of passing by the closest. The equality of all is borne by my inequality, the surplus of my duties over my rights. The forgetting of self moves justice.
>
> (Levinas, 1981, p. 159)

In addition, philosopher J. Aaron Simmons (2012) argues convincingly that we face what he calls a "metaethical emergency" (p. 229), a situation like that facing Britain in 1939 as it considered whether the ordinary laws of war still applied in facing an imminent threat to all civilized life. That we confront, Simmons claims, the looming destruction of a livable world means we do not have the leisure for non-anthropocentric ethics concerned equally for all species. We must, he believes, immediately prioritize the most vulnerable humans suffering from the effects of climate change if we are to have any chance of motivating the type, extent, and rapidity of change needed. Ordinary ethics, such as the duty and utilitarian theories outlined above, even including deep ecology perhaps, belong to everyday life. "In the context of a metaethical emergency," writes Simmons, "there is no time to advocate theories that do not have a good enough chance of motivating a reasonable public response to the crisis" (p. 232). We have to treat our current situation as an extreme humanitarian emergency, not something to measure against competing ideals and philosophies. Paraphrasing Simmons, emergencies call for a kind of ethical triage that he calls "a hierarchy of ethico-political significance" (p. 230). William Edelglass (2012), in the same vein, writes that "Levinas provides a way of understanding how the singular subject's moral responsibility is the condition for the possibility of collective responsibility" (p. 211). Response to the face of the suffering sister and brother creates political change through prophetic action (Anderson, 2006; Orange, 2016).

Remembering the biblical story, we notice a missing shame in Cain's insolent response: I do not know: am I my brother's keeper? Without mentioning Cain's murderous violence, the question, "Where is Abel your brother?" comes to him as a gracious invitation to repentance and repair. Cain shamelessly refuses its underlying premise of responsibility, and models not only the "we knew nothing of the extermination camps," but also the "send them back where they came from—this is our country." If they aren't smart enough to get rich from oil, fracking, and coal, too bad for them. In the face of such attitudes, rampant in Europe and North America, only a robust solidarity, enacting our answer to Cain, can begin to solve the problems of climate justice. We can make common cause with romantics who love the earth and its many species—I do myself, as a native Oregonian—but fundamentally we must find ourselves as our other's keepers.

We can remember here also the African ethics of *Ubuntu*, well-explained by recently transitioned Archbishop Desmond Tutu:

> A person with Ubuntu is open and available to others, affirming of others, does not feel threatened that others are able and good... knowing that he or she belongs in a greater whole and is diminished when others are humiliated or diminished, when others are tortured or oppressed.
>
> (Tutu, 1999: 31)

Conclusions

First, what can we not conclude? Clearly the radical ethics we have been discussing cannot help either psychotherapists or international public policy experts to decide the best road to climate justice: "equal per capita entitlements, rights to subsistence emissions, priority to the least well off, or equalizing marginal costs" (Edelglass, 2012, p. 227; Gardiner, 2011). Nor does radical ethics tell us how to move the seemingly unmovable forces of power and money arrayed against climate justice. These same power-systems forbid us to associate climate change and its effects with social justice. We must look to the wisdom accumulated by those who overturned apartheid, organized the civil rights movement, and finally brought the Vietnam war to a close, as well as to our partners in the indigenous communities so affected by climate devastation, to find our way. Radical ethics does not provide rules, strategies, political philosophy, and structures. Instead, it confronts each and all of us with our responsibility, without beginning and without end, inescapable, for our world and for each other.

Radical ethics means that we cannot go on as we did yesterday, self-satisfied that we are doing our best, or shifting our personal responsibility (Edelglass, 2012) onto "the system." The terrified faces of the destitute refugees, of those whose homes are being turned into desert or going beneath the sea, threatened by violence, forbid me to sleep comfortably and command me to respond. Every day I must allow them to persecute me, to pull me out of my comfortable life, to make me non-indifferent. For each of us, response will take its own form, depending on how and where we see the useless suffering and hear the cries, and on what our own health allows.

And yet, a radical ethics never begins and never ends. The face of the suffering other demands response from me before any agreement ever made (yes, I am my other's keeper), and this responsibility continues. "I more than all the others," as both Dostoevsky and Levinas repeatedly wrote. For the extreme situation in which, once again, we find ourselves, perhaps only the ethic of hyperbolic responsibility can shake us out of our complacency. If not to save our own skin, still comfortable enough, am I not my sister's and brother's keeper?

And what if we have already procrastinated too long? Watching Greta Thunberg, I want to believe we can all wake up and respond. But looking at Paradise, California and Mozambique and the Missouri River, who knows? Is it time to mourn, to accept responsibility for the ethically unconscious way our generation has lived—after all, half the greenhouse gasses in our atmospheres and oceans have come in the past 25 years? Do we retreat to higher ground, leaving the poor to drown and starve? To use the language of Hans Loewald, how do we atone not only for parricide but for infanticide and for fratricide? How do we live as humanly as possible in the time that is left? I leave you with such questions.

Notes

1 At the end of World War II a similar task confronted Germans who could see only the rubble and their own suffering, not its causes brought on by their own crimes (cf. Frie, 2016).
2 I am grateful to my neighbor Jack Jackson, a Seneca, for reminding me of this preference, one I had heard strongly voiced by Hopis too.
3 Writing of both Levinas and of Jean Laplanche, she continues: "To understand the unconscious… is to understand what *cannot* belong, properly speaking, to me, precisely because it defies the rhetoric of belonging, is a way of being dispossessed through the address of the other from the start" (p. 54).
4 In a thought-provoking footnote, James Hatley (2005) writes: "The German term, *Verantwortung,* seems preferable here to the English word 'responsibility.' The German word hints at an intensification of response to the point of being utterly taken up in answering what calls one, whereas the English term's etymology suggests that the ability to respond limits the call to respond" (p. 51, n14). "Murder," writes Hatley about Cain in an earlier footnote (pp. 50–51, n11) "is an attempt to escape the address of the other, to render the world as faceless."
5 Granted that Levinas himself failed to speak out against the violence at Sabra and Shatila in 1982, and/or to see his own Eurocentrism and masculinism Critchley (2004).

References

Anderson, K. (2006). Public Transgressions: Levinas and Arendt. In A. Horowitz & G. Horowitz (Eds.), *Difficult Justice: Commentaries on Levinas and Politics* (pp. 127–147). Toronto: University of Toronto Press.

Bernasconi, R. (2006). Strangers and Slaves in the Land of Egypt: Levinas and the Politics of Otherness. In A. Horowitz & G. Horowitz (Eds.), *Difficult Justice: Commentaries on Levinas and Politics* (pp. 246–261). Toronto: University of Toronto Press.

Bernasconi, R. (2010). Globalization and World Hunger: Kant and Levinas. In P. Atterton & M. Calarco (Eds.), *Radicalizing Levinas* (pp. 69–86). Albany, NY: State University of New York Press.

Blackmon, D.A. (2009). *Slavery by Another Name: The Re-enslavement of Black Americans from the Civil War to World War II.* New York: Anchor Books.

Brenner, C. (1962). Freud's Concept of Repression and Defense, Its Theoretical and Observational Language. *Psychoanalytic Quarterly,* 31: 562–563.

Bromberg, P. (2006). *Awakening the Dreamer: Clinical Journeys.* Hillsdale, NJ: The Analytic Press.

Butler, J. (2004). *Precarious Life: The Powers of Mourning and Violence.* London, New York: Verso.

Butler, J. (2005). *Giving an Account of Oneself*. New York: Fordham University Press.

Butler, J. (2010). *Frames of War: When Is Life Grievable?* London: Verso.

Chefetz, R. (2000). Disorder in the Therapist's View of the Self. *Psychoanalytic Inquiry*, 20(2): 305–329.

Coates, T-N. (2015). *Between the World and Me*. New York: Spiegel and Grau.

Critchley, S. (2004). *Very Little—Almost Nothing: Death, Philosophy, Literature*. London, New York: Routledge.

Cushman, P. (2007). A Burning World, an Absent God: Midrash, Hermeneutics, and Psychoanalysis. *Contemporary Psychoanalysis*, 43: 47–88.

Davis, D.B. (2006). *Inhuman Bondage: The Rise and Fall of Slavery in the New World*. London: Verso.

Edelglass, W. (2012). Rethinking Responsibility in an Age of Anthropogenic Climate Catastrophe. In W. Edelglass, J. Hatley, & C. Diehm (Eds.), *Facing Nature: Levinas and Environmental Thought* (pp. 209–228). Pittsburgh, PA: Duquesne University Press.

Freeman, M.P. (2012). The Narrative Unconscious. *Contemporary Psychoanalysis*, 48(3): 344–366.

Frie, R. (2016). *Not in My Family: German Memory and the Holocaust*. Oxford: Oxford University Press.

Gardiner, S. (2011). Climate Justice. In J.S. Dryzek, R.B. Norgaard, & D. Schlosberg (Eds.), *The Oxford Handbook of Climate Change and Society* (pp. 309–322). Oxford: The Oxford University Press.

Grosskurth, P. (1986). *Melanie Klein: Her World and Her Work*. New York: Knopf.

Hatley, J. (2005). Beyond Outrage: The Delirium or Responsibiity in Levinas's Scene of Persecution. In E. Nelson, A. Kapust and K. Still (Eds.), *Addressing Levinas* (pp. 34–51). Evanston, IL: Northwestern University Press.

Herman, J.L. (2009). Crime and Memory. In K. Golden & B. Bergo (Eds.), *The Trauma Controversy: Philosophical and Interdisciplinary Dialogues* (pp. 127–141). Albany, NY: State University of New York Press.

Herman, J.L. (2011). Posttraumatic Stress Disorder as a Shame Disorder. In R. Dearing & J. Tangney (Eds.), *Shame in the Therapy Hour* (pp. 261–276). Washington, DC: American Psychological Association.

Howell, E.F., & Itzkowitz, S. (2016). *The Dissociative Mind in Psychoanalysis: Understanding and Working with Trauma*. Abingdon, New York: Routledge.

Jacobson, E. (1957). Denial and Repression. *Journal of the American Psychoanalytic Association*, 5: 61–92.

King, M.L. (1967). I Have a Dream. In *The Essential Writings of Martin Luther King, 1986*. San Francisco, CA: Harper Collins.

Kohut, T. (2003). Psychoanalysis as Psychohistory or Why Psychotherapists Cannot Afford to Ignore Culture. *Annual Psychoanalysis*, 31: 225–236.

Kohut, T. (2012). *A German Generation: An Experiential History of the 20th Century*. New Haven, CT: Yale University Press.

Kuriloff, E. (2014). *Contemporary Psychoanalysis and the Legacy of the Third Reich: History, Memory, Tradition*. New York: Routledge, Taylor & Francis Group.

Lertzman, R. (2015). *Environmental Melancholia: Psychoanalytic Dimensions of Engagement*. New York: Routledge.

Levinas, E. (1969). *Totality and Infinity: An Essay on Exteriority*. Trans. A. Lingis. Pittsburgh, PA: Duquesne University Press.

Levinas, E. (1981). *Otherwise Than Being: Or, Beyond Essence*. The Hague, Boston Hingham, MA: M. Nijhoff; Distributors for the U.S. and Canada, Kluwer Boston.

Loewald, H.W. (1955). Hypnoid State, Repression, Abreaction and Recollection. *Journal of the American Psychoanalytic Association*, 3: 201–210.

Marshall, G. (2014). *Don't Even Think About It: Why Our Brains Are Wired to Ignore Climate Change*. New York: Bloomsbury.

McKibben, B. (2014). *Oil and Honey: The Education of an Unlikely Activist*. New York: St. Martin's Griffin.

Orange, D. (2016). *Nourishing the Inner Life of Clinicians and Humanitarians: The Ethical Turn in Psychoanalysis*. London, New York: Routledge.

Ryde, J. (2010). Supervising psychotherapists who work with asylum seekers and refugees: A space to reflect where feelings are unbearable. In R. Shohet (Ed.), *Supervision as Transformation: A Passion for Learning*. London: JKP.

Simmons, J.A. (2012). A Relational Model of Anthopocentrism: A Levinasian Approach to the Ethics of Climate Change. In W. Edelglass, J. Hatley, & C. Diehm (Eds.), *Facing Nature: Levinas and Environmental Thought* (pp. 229–252). Pittsburgh, PA: Duquesne University Press.

Smith, M. (2011). Ecological Community, the Sense of the World, and Senseless Extinction. *Environmental Humanities*, 2: 21–41.

Tutu, D. (1999). *No Future Without Forgiveness* (1st ed.). New York: Doubleday.

Volkan, V.D. (1994). Repression and Dissociation: Implications for Personality Theory, Psychopathology, and Health. *Journal of the American Psychoanalytic Association*, 42: 301–304.

Westphal, M. (2008). *Levinas and Kierkegaard in Dialogue*. Bloomington, IN: Indiana University Press.

Chapter 7

The Thought of the Desert and the Desert of Thought

Michael Robert Kelly and Brian R. Clack

Deserts call us, teach us. They challenge us mentally, physically, and spiritually. And from them we learn about being and our human being if we can listen in silent reverence. Mary Hunter Austin dedicates her 1903 writings about California's Death Valley and Mojave Desert, *The Land of Little Rain*, to "Eve," with the inscription, "The comfortress of unsuccess." Such is with the desert, too, the celebrant of non-instrumental value, of intrinsic value, of the value of the valueless, of the sublime, the wonderful, the strange, the eerie, the mirage, and so on. Austin described these California deserts as "a land of lost rivers with little in it to love; yet a land that once visited must be come back to inevitably. If it were not so there would be little told of it."[1] Perhaps it is the desert that hosts us rather than we who host the desert? Such is our wager in what follows.

The desert always gets to one. It gets to one because it gets one thinking that one does not know what to think. One may go to the desert out of curiosity or when desiring reprieve from urban and suburban social-political life—to detoxify by escapist diversion in nature's sauna. And it helps. Until, that is, one really leaves the hotel and actually hits the desert. Once there, one really doesn't know how or what to think—neither about the desert nor its effect on oneself. The desert leaves the most articulate and learned among us floundering, reducing us to really bad poets or those encyclopedists that know what the desert is in one sense but still manage to miss its relation to human meaning and thus understand nothing of it in another sense.

It may be that our inarticulate floundering results from the confrontation of our human expectations with the profound indifference of the markedly inhuman desert. For the desert confounds that characteristic human need to understand and utilize. Edward Abbey, among the very finest champions of the desert wilderness, experienced it as "spare, sparse, austere, utterly worthless, … a realm beyond the human."[2] Thinking about such a realm requires an adjustment of the human, *away* indeed from the human. We seem unable to impose our habitual state of being and thinking here, an insight memorably voiced by Barbara Kingsolver: "It did not take me long in the desert to realize I was thinking like a person, and on that score was deeply outnumbered."[3] Some kind of adjustment or even reversal of the pattern of thought seems called for, and in a perceptive observation on Abbey's retreat to the desert, Robert Macfarlane remarks (in a way that harkens to Maurice

DOI: 10.4324/9781003456940-10

Merleau-Ponty's notion of the *chiasm*) that "he goes to nature in order not just to think about nature but also to think *with* nature, and even more radically to be thought *by* nature."[4] The upshot of this must be some degree of wariness toward any attempt at understanding the desert realm.

Sure, anyone trained in understanding some feature of the desert and its relevance to human meaning—for example, the scientist, religious studies scholar, anthropologist, park ranger, and so on—can provide information about the desert and its natural, cultural, historical, symbolic value. But each partial presentation of the whole can only go so far; such *information* is not what we are after when we find that the desert is after us.

So inclined are we to seek further (and usually explanatory) information when confronted by a phenomenon that is puzzling or impressive that it is hard to imagine that something different may be needed. One such different path may be detected in Wittgenstein's response to Sir James Frazer's anthropological enquiries. A reader of the *Remarks on Frazer's Golden Bough* will be struck by Wittgenstein's insistence that *any* attempt to find an explanation of some piece of ritual behavior (say, human sacrifice) is *wrong*; and that what is required instead is an activity—"putting together in the right way what we *know*"—resulting in something variously described as "satisfaction," "peace," and "solace."[5] We can note how non-epistemic those goals are. For Wittgenstein, then, *understanding* ritual phenomena is distinct from finding historical or psychological *explanations* for such phenomena, and is more a matter of *coming to terms* with something that disquiets or troubles us. Might something comparable be true of our engagement with the desert?

We can understand loads of natural, cultural, historical, and symbolic facts about the desert without understanding a single thing about why the desert provokes human imagination, awe, wonder, perplexity, and confusion. When we visit the desert, we do not inhabit the scientific desert, and we no longer inhabit the world of those who really contended with the desert—the ancient Babylonians and Israelites, the explorers, Mary Hunter Austin, etc. Science surely can explain the desert's material composition and function, can demythologize and show to be incorrect ordinary beliefs about this supposedly barren and lifeless place. But that constitutes an absurdity in the very sense of deaf, dull, out of tune. It is not what the host or *sojourner* is after when visiting the desert. For the scientist *as* scientist, scientific meaning about the desert provides meaning; for the scientist *as* desert host (hosting the desert or hosted by the desert in a kind of chiasmic *Ineinander*) the scientific meaning provides no meaning. As to what the scholarly social sciences can tell us of the desert, we can come to know—be informed about—how the desert factored into various cultures and cultural meanings in various places across various times in human history. But we do not inhabit *that* meaning. We *cannot* inhabit that meaning in this strange, hosting-hosted relationship one finds in the desert. We are left with our own bad poetry and reaching contemplations.

We might feel that the category of the sublime may do some work for us here, that it may help us to understand and appreciate the "weird solitude, the great silence, the grim desolation" of this forbidding landscape that confronts us.[6] And it is certainly the case that the idea of the sublime, particularly as it is characterized by Burke, seems especially fitted to our experience of the desert, an experience with that uneasy mix of wonder and admiration, terror and delight, condensed into what Addison called "an agreeable kind of horror."[7] Those sublime qualities of vastness, privation, obscurity, roughness, and the painfully bright light of the desert sun, all contribute to our astonished sense of being insignificant and overwhelmed by enormous and indifferent power. Do we not here "shrink," as Burke puts it, "into the minuteness of our own nature, and are, in a manner, annihilated"?[8] But here one must too be careful. We must not too readily assume that a mere arrangement of words and concepts can compass the world of the desert; too often indeed that schema of the sublime can become tired out, devalued by overuse and laziness; we say "That is sublime," and then move on. A temptation to be resisted.

Might the thorough avoidance of words be a wise strategy? Sympathetic to thinking of the desert as sublime, and indeed as "awful" (a far better word than the now-impotent and worn-out "awesome"), Joseph Wood Krutch stressed its distinctive voice and mood: "For one thing," he writes, "the desert is conservative, not radical."[9] What might it mean to see the desert as conservative? It may highlight the very slow and silent changes that happen in it. Krutch knew his Samuel Johnson well, having written a biography of him;[10] and Johnson, being the good Tory that he was, spoke often of his antipathy to radical and ambitious political projects, stressing instead that the "progress of reformation is gradual and silent, as the extension of evening shadows."[11] And the desert may, at least for some of its admirers, be suggestive of gradual, imperceptible change, if not of downright immutability. It seems solid and immoveable, unaffected by the chances and changes and vicissitudes that bedevil our fugitive human lives. Note Johnson's stress on silence in his gradualist ideal of reformation. Silence had been a notable feature of the European conservative tradition, Adolf Grabowsky defining the conservative attitude as a "silent reverence for the impenetrable."[12] Might these words be as good as any for capturing the profound feeling that grips us when we find ourselves in awful desert landscapes? We cannot grasp what surrounds us, we cannot fathom it. Words fail us. In the extremity of the desert, we find ourselves, in Belden Lane's formulation, "running out of language, driven to silence."[13] *Silent reverence for the impenetrable.*

The desert confronted; the desert thought beyond its natural elements or cultural meanings no longer inhabitable, leaves human thought deserted like that biggest of questions do: Why is there something rather than nothing? And why is there *this* kind of something rather than nothing? No matter how much sense we can impart to the desert via these various natural and social sciences, the desert remains before us as a stark reminder of *what* we cannot say. It's being makes no sense to us—zero—if we try to think it rather than reverently learn something of our human selves from it. We should admit this claim or at the very least entertain it very seriously. Albert Camus does so unflinchingly.

In *The Myth of Sisyphus and Other Essays*, Camus presents a desert refrain. With regret, he notes,

> There are no more deserts… Yet there is a need for them. In order to understand the world, one has to turn away from it on occasion; in order to serve men better, one has to hold them at a distance for a time. But where can one find the solitude necessary to vigor, the deep breath in which the mind collects itself and courage gauges its strength?

To that question, Camus offers a suggestive response of a rather broad nature:

> The desert itself has assumed significance… For all the world's sorrows it is a hallowed spot. But at certain moments the heart wants nothing so much as spots devoid of poetry. Descartes, planning to meditate, chose his desert: the most mercantile city of his era. There he found his solitude and the occasion for perhaps the greatest of our virile poems: "The first [precept] was never to accept anything as true unless I knew it to be obviously so." It is possible to have less ambition and the same nostalgia… In order to flee poetry and yet recapture the peace of stones, other deserts are needed, other spots without soul and without reprieve.[14]

Despite these remarks, neither Descartes nor the skeptics to whom he responds entered the desert—or the desert of thought as Camus puts it elsewhere—because neither faced the absurd. And to outline what Camus dislikes about Descartes' thought may help us grasp what Camus means by the absurd, which is the same as grasping what Camus means by the importance of deserts and the desert of thought to which he encourages us to turn.

Camus oddly characterizes Descartes' philosophical method and ambition—which we must recall given the desert's own resistance to utility value and Descartes' aim that we become "masters and possessors of nature"—as a great poem.[15] Camus differs dramatically. Camus encourages "less ambition [but] the same nostalgia as Descartes," and he encourages us—after having marked Descartes' *Meditations* as "perhaps the greatest of our virile poems"—"to flee poetry" in order to "recapture the peace of stones." On the cusp of facing the realization that it is impossible to have any certain knowledge about anything whatsoever in the world, in science, and in the universe, Descartes, on one reading of Camus, fled the desert of thought rather than entering it. Using his power of reason, Descartes finds a reprieve from skepticism by arriving via the road of the skeptical method at an indubitable truth in the mode of the impossibility of doubting one's own existence. In his supposed desert of thought, Descartes find certainty in his own existence as a thinking thing or a thing that thinks—a disembodied soul. Pure life, pure sense, pure animus—no desert. Descartes did not engage the absurd, did not host the desert; he tried to silence it, to master and possess it. In doing so, Descartes provided Western culture a reprieve from skepticism and a foundation for the new science and technology but left it deserted—a disembodied soul—without deserts. Camus wagers, however, that we need places "without soul and without reprieve," that is, deserts.

Camus diametrically opposes the absurd to Descartes's conviction regarding the human mind's ability to arrive at certain knowledge impervious to skepticism. In such a flight from the desert of thought, Descartes's poetry did not—could not—manifest the absurd, i.e., the admission that, as Camus puts it, "the world in itself is not reasonable [and] that is all that can be said."[16]

Enter the desert, that chiasmic *Ineinander*: "The absurd depends as much on man as on the world."[17] Camus describes the absurd as the confrontation of the arational universe with the innate human longing for explanation, understanding, and clarity—"the wild longing for clarity whose call echoes in the human heart" (the aforementioned nostalgia).[18] The absurd human being—a willing desert host unlike Descartes—concerns herself with not explaining and solving but experiencing and describing. Such is where the desert leaves the honest agent of contemplation. Alluding to the desert of thought in which one might make manifest the absurd—that is, the going forward in the face of the moment when one realizes that the universe doesn't give a damn about human being and its wild longing for understanding, order, and meaning—Camus writes, "The method defined here acknowledges the feeling that all true knowledge is impossible. Solely appearances can be enumerated and the *climate* makes itself felt" (p. 12). Feeling, climate, atmosphere, mood: This is neither a natural or social science nor a philosophy—nor poetry as we have seen.

Camus doesn't mean that the knowledge that we can gain via science and social science is valueless. Camus wasn't naïve or anti-science or anti-reason. But he was anti-scientism and anti-rationalism. And he surely didn't think we should equate knowledge full stop with the knowledge provided by science or rational calculation. Camus' repetition of the concept of climate harkens to a mood—an atmosphere—a feeling or an affective encounter of oneself—of human being—with the universe in all its indifferent silence (as Abbey notes of the desert itself).[19] As Camus notes of absurd or the desert of thought, it "sums itself up as a lucid invitation to live and to create in the very midst of the desert," that is, to live and create in the midst of the confrontation between the human longing to know and the arational universe that remains silent in response to our queries regarding the meaning of it that reason, science, history, culture, and symbolic demonstration alone cannot capture. The method that Camus's desert of thought advises is not only one that "acknowledges the *feeling* that all true knowledge is impossible," but it is also one that speculates that "solely appearances can be enumerated."

Even when reading Camus we find we need more deserts.

The desert itself instills in us the wonder of the useless and existence and its various and sometimes strange phenomena. Let us play around with the notion of the desert of thought to "enumerate appearances."

To grasp the importance of desert thoughts and thus grasp better the desert of thought, we perhaps should reject the entailments around Camus' idea that "solely appearances can be enumerated"—as if appearances were but mere appearances and thus somehow deficient modes of apprehending a reality that hides always behind an appearance. Like Camus, we want to avoid the pitfalls of rationalism and scientism on the one hand and methodological skepticism on the other hand. Unlike Camus, however, we do not want the desert of thought to metaphorically convey the standpoint of the curious mind confronting in heroic defiance the universe's silent refusal of our deepest wishes: Even if we must be on guard against scientism and rationalism, appearances are not mere appearances.[20]

Consider the paradigmatic desert phenomenon: the mirage. The desert has its mirage, the cloud its halo, the water droplets its rainbow. Of these three, only the mirage is thought as nothing—or no real thing—but a mere appearance, a mere illusion as if an illusion were but an impoverished appearance rather than the distinctive phenomenon that it is. Whereas we tend to demean the mirage, we marvel at the rainbow and the halo as legitimate (not illusory) phenomena—a fleeting moment of beauty that calls our attention to the extraordinary so much so that the one lucky to first behold it harkens others—even strangers—to join attention and marvel along. The rainbow and the halo each mark one of life's beauties that fills the moment in time by fulfilling the fortunate viewer momentarily.

The mirage, however, is often thought not to fulfill but to disappoint. Common sense takes a mirage as a mere appearance or something that it believes to be nothing (but an illusion). Science tells common sense that is not quite correct. Science tells us that the mirage is a nothing that is a something atmospherically speaking. But each takes the mirage as a *mere* appearance. A mirage, however, is no mere illusion. A mirage is a wonder, a miracle.

In visual or optical illusions, our mind "sees" what it expects to see rather than what our eyes really see. A mirage, by contrast, is a physical phenomenon that happens *before* us. When one sees a mirage, one sees *exactly* what is there to be seen. There's no trick of perception or incongruity between what is presented on the side of the world and what is apprehended on the side of the perceiver. That is what an illusion is that a mirage is not, namely, an incongruity between what is presented on the side of the world and what is apprehended on the side of the perceiver (hence precisely the magic of the illusionist or magician). A mirage is not an illusion at all, for when we see a mirage we perceive something that is really there: the intersection of really hot ground and much cooler air. The mirage *really* is there. It is a real phenomenon and neither a mere appearance nor a visual trick. The illusion is deceptive. The mirage is miraculous—as the etymology of the term conveys: from Latin *mirus*, "wonderful." Moving further against misunderstanding the mirage, recall that its Latin root shared a sense of miracle, as "*mirari*," which conveyed the sense "to look at as in admire or wonder." The desert host reminds us of the many wonders of existence that themselves motivate the desert of thought.[21]

What, then, of the desert of thought motivated by the thought of the desert? Does the phrase mean a thought not alive? No longer alive or a yet to be born? Maybe neither since all that is alive or has a soul exists although not all that exists has a soul or is alive (as surely the desert reminds us). Does the desert of thought mean an abandoned or deserted thought?

Notice the different temporal dimensions that each hypothetical interpretive option entails. Does each not have its own way of being or appearing that is no mere appearance even if each moment of the manifold of appearances does not claim to be the whole of being? Why conclude that the past no longer exists? We believe that unperceived places and spaces exist. Does the past not survive? Perhaps the past survives—like the desert—seemingly fixed and unchanged? Are we deserted by the past or do we desert the past? Consider the repetition of ruminations or obsessive thoughts like the desert's haunting seams.

Refrain from overquoting Faulkner.

Become a bad poet.

Maybe the desert of thought expresses that thought that—switching from the past to the present—surprises us—as deserts, too, do. Deserts surprise us such as when they show life unexpected. Latent vegetation may spring ephemerally; a desert reptile may crawl upon us like a harsh thought that slithers into one's consciousness unforeseen and startling. All surprise, pleasant or unpleasant, good or bad, presents in the present. And yet again, perhaps the thought of a desert of thought as a thought not yet born suggests hope, turning our minds to the futural mode of appearing of the desert of thought. And there is nothing dead or deadening in hope, although in hoping we remain in the domain of the inexistent with respect at least to that for which we hope. Might hope reside in the desert? Another wonderful paradox to consider—hope symbolized in the desert.

Whatever the desert of thought may mean, the metaphor perhaps stems from and reflects something of the desert host that motivated Camus' phrase. As Lane puts it,

> The desert demands the last word—even though it remains a silent place of forbidding mystery, absorbing the sound of all other voices in its fierce terrain… a kind of Greek Chorus never speaking but always present, offering its own critique of everything else that's said, silently deconstructing every naïve and romantic notion.[22]

The desert host teaches at the very least that silent reverence for the impenetrable.

Notes

1 M. Austin, *The Land of Little Rain* (Boston and New York: Houghton, Miflin, and Company, 1903), p. 6.

2 E. Abbey, *Desert Solitaire* (London: William Collins, [1968] 2020), p. 289.

3 B. Kingsolver, *High Tide in Tucson* (New York: HarperCollins, 1995), p. 32.

4 R. Macfarlane, "Introduction" to E. Abbey, *Desert Solitaire*, p. 5.

5 L. Wittgenstein, *Remarks on Frazer's Golden Bough* (Retford: The Brynmill Press, 1979), pp. 2, 3. Note in particular his remark that "for someone broken up by love an explanatory hypothesis won't help much.—It will not bring peace [*beruhigen*]" (p. 3).

6 J.C. Van Dyke, *The Desert* (New York: Charles Scribner's Sons, 1901), p. 19.

7 J. Addison, *Remarks on Several Parts of Italy* (London: J. Tonson, 1733), p. 261.

8 E. Burke, *A Philosophical Enquiry into the Origin of Our Ideas of the Sublime and Beautiful* (London: F. & C. Rivington, 1798 [1756]), p. 119.

9 J.W. Krutch, *The Voice of the Desert* (New York: William Sloane Associates, 1954), p. 221.

10 J.W. Krutch, *Samuel Johnson: A Biography* (New York: Henry Holt and Company, 1944).

11 S. Johnson, *The Adventurer*, 1754, in *The Works of Samuel Johnson* (London: George Cowie & Co., 1825), Vol. II, p. 137.

12 A. Grabowsky, quoted in J.C. Nyíri, "Wittgenstein's Later Work in Relation to Conservatism," in B. McGuinness (ed.), *Wittgenstein and His Times* (Oxford: Basil Blackwell, 1982), p. 56.

13 B.C. Lane, *The Solace of Fierce Landscapes* (New York: Oxford University Press, 1998), p. 39. Note also Lane's final judgment: "The desert is where one confronts one's inevitable loss of control, the inadequacy of language, the spectre of one's own demise" (p. 231).

14 A. Camus, "The Minotaur *or* The Stop in Oran," in *The Myth of Sisyphus and Other Essays*, trans. J. O'Brien (New York: Vintage Books, 1955), pp. 157–159.

15 R. Descartes, *Discourse on Method*, trans. I. Maclean (New York: Oxford, 2006), p. 51.

16 ibid., p. 21.

17 ibid.

18 ibid.

19 E. Abbey, *Desert Solitaire*, p. 235: "Alone in the silence, I understand for a moment the dread which many feel in the presence of the primeval desert, the unconscious fear which compels them to tame, alter or destroy what they cannot understand, to reduce the wild and pre-human to human dimensions. Anything rather than confront directly the ante-human, that *other world* which frightens not through danger or hostility but in something far worse—its implacable indifference."

20 R. Sokolowski, *Introduction to Phenomenology* (New York: Cambridge University Press, 2000), p. 8.

21 "mirage, n." OED Online. March 2023. Oxford University Press. www-oed-com.sand-iego.idm.oclc.org/view/Entry/119067?rskey=EYOUNe&result=1&isAdvanced=false (accessed April 21, 2023).

22 B.C. Lane, *The Solace of Fierce Landscapes*, p. 231.

Chapter 8

Cosmological Persons: Bringing Healing Down to Earth

Chandler D. Rogers

Almost all who write on the climate emergency are rightly critical of the hubris that has led our species to become a particularly parasitic guest to the earth, our host. But the answer to human hubris cannot simply be to emphasize that our species is no different from any other. This too easily becomes an excuse to perpetuate the harms that we are already inflicting. We *are* different. We are uniquely responsible for the climate emergency, and to clean up the mess we have made. That mess, I contend, is a *personal* mess. That emergency is a personal emergency. As such, ecological healing would require addressing directly three all-too-human tendencies that have been exacerbated in the modern period.

As persons we are irreducibly unique subjects, existing essentially in relation to one another. But largely obscured in the more recent history of Western thought is the *cosmological dimension* of our personhood: our existence in relation not just to other humans, but also to other animals, plants, and ecosystems. This is the sense in which we, as persons, exist as part of the larger cosmos, conceived as a meaningful whole. To reconnect to this dimension of our personhood we must learn to resist the anesthetizing, alienating, and anonymizing tendencies that have characterized the recent history of our species.

In what follows I propose three remedies to these three kinds of sickness. We need to re-cultivate sensitivity toward, to get back in touch with, and to speak out on behalf of the wild plants, animals, and ecosystems that have been forced to the periphery of human society. Yet the aim of our critique cannot be mere disdain for techno-scientific control and the domination of anonymous market forces.[1] Our goal must be twofold: to regain touch with our animality by embracing the earth that hosts us, thus becoming more gracious guests, and to become more fully human by becoming better hosts to the multitude of earth's creatures, whose lives hang in the balance.

Anesthetization

Anesthetization is the state of having become insensitive to pain or suffering, and more broadly to that which stimulates the senses. It includes reference also to a cultural crisis in aesthetics, or to the modern victory of *scientific* knowing—and the objectification of all that is not mind—over *embodied* ways of knowing. The latter

DOI: 10.4324/9781003456940-11

are in fact integral components of a largely absent complement to scientific objectification: tactful philosophical knowing. Reintegration of the latter would require a healthy dose of aesthetic sensibility, or in other words *aestheticization*.[2]

We have become anesthetized to the living world that beckons to us from beyond our societal constructs. From the onset our embodied responses are solicited through the heightening of our senses, but the affective primacy of touch has been gradually eclipsed by the abstract blessedness of sight (Kearney 2021, 33f; Treanor 2023, 19f; Fitzpatrick 2023, 35). An aesthetic way of being, in which a subject connects meaningfully to its surrounding world, has been displaced by the objectivizing gaze: "Aesthetics is a way of being, a stance toward the world; an aesthetic experience requires a relationship between a seeking subject and a responsive world. But scenery is a stockpile of usable commodities" (Evernden 1999, 54). One byproduct of the shift toward objectivistic knowing is that plants, animals, and their surrounding biotic environments now blend into a scene to be viewed from above and outside. Earth's wild places become spectacles to be viewed on occasion, rather than a living theater inviting us to participation (cf. Merleau-Ponty 2012, lxxii).

Aristotle does carefully attend to touch, by contrast with Plato. But like Plato, he also famously argues, at the opening of his *Metaphysics*, that sight is the most blessed of the senses. Sight takes in the scene all at once, as touch struggles to grasp the whole. But while blessed rationality strives, perhaps inevitably, toward an ultimate perspective on things, our human limitations prevent us from ever wholly attaining to the God's-eye-view. The great danger has been that when we come close to such a blessed standpoint, our own animality becomes utterly neglected or repressed; or perhaps falsely claimed to have been overcome. The wonders of touch, in its tactful attention to detail, and particularity, are overlooked (cf. Kearney 2021, 36).

Modern science in particular hypostatizes sight over and against touch, and prizes objectification over value-laden knowing. But perhaps this situation presents in part a necessary predicament. The phenomenologist Max Scheler, in *On the Eternal in Man*, maintains that scientific knowing is contingent upon the following moral precondition: "*self-mastery must be achieved:* in this way it is possible to objectify the instinctual impulses of life, which are "given" and experienced as "of the flesh" and which must needs exert a constant influence on natural sensory perception" (Scheler 2010, 93).[3] Scheler suggests that one's scientific knowledge will be more or less adequate in proportion to the extent of one's mastery over instinctual impulses and desires; that objective knowledge of things will correspond to the degree to which the scientist has been able to suspend subjective desires and intentions. But after centuries of practice, those modes of subjectivity have been effectively buried under what the ecophenomenologist Erazim Kohák calls a "heavy layer of forgetting" (Kohák 1987, x). In the face of significant disagreement about values—whether amidst the cultural upheavals of the seventeenth century, or in the context of increasing globalization, and techno-democratic assertions toward individual authority, in the twenty-first—scientific objectivity has taken absolute epistemic precedence, thus overriding the meaning-laden dimension of more subject-dependent insights concerning value, empathy, and virtue.

If self-mastery can be defined as "the domination of the instinctual impulses by the rational will," then in acts of scientific knowing the possibility of instrumentalization is already implicit: the possibility of manipulating and controlling the entity or entities in question for the sake of some desired end (Scheler 2010, 96). Without further moral qualification, the scientific knower is especially tempted toward the manipulation of the objects of his knowledge "by means of a rational will which is itself guided, but also *bound*, by possible anthropocentric aims and values" (Scheler 2010, 96–97).[4] Scheler contrasts scientific knowing with its missing counterpart, philosophical knowing:

> [I]t is not by chance but in the nature of things that even the scientist's basic *moral* attitude to the world and his task in it is, and should be, totally different from that of the philosopher. In positive research the scientist's will to know is primarily inspired by a will to *master* and, thence arising, a will to *order* the whole of nature: it is for that very reason that "laws," in *obedience* to which nature lets herself be governed, represent the highest goal of his endeavour. What interests him is not *what* the world is, but how it may be considered as *constructed*, so that, within the scope of this highest goal, it may be regarded as practically *modifiable*.
>
> (Scheler 2010, 97)

Scientific knowing must be counterbalanced by the wisdom of the more holistic approach offered by philosophical knowing, which occurs also only on the condition of the kind of moral self-mastery described above.[5] Tempering the will to control and modify, however, are two additional moral preconditions. First, in philosophical knowing "the whole spiritual person must love absolute value and being." Second, "the natural self and ego must be humbled" (Scheler 2010, 95; emphasis removed). It is by way of these two requirements, love of absolute value and being and humbling of the natural self or ego, that the philosopher's way of knowing can temper the unbridled and exalted techno-scientific will to mastery.[6]

An aestheticized love of wisdom which remains faithful to the earth, in the words of Nietzsche's *Zarathustra*, will resist the otherworldly drive currently fueling civilizational pursuits toward an endless excess (Nietzsche 2005, 12). Renewed aesthetic sensibility can help to introduce a healthy dose of humility into our cultural projects, with an eye to reversing the markedly noxious results of our more recent tenure as parasitic guests of the earth.[7]

Alienation

Alienation is a state of estrangement from the earth, from animality, and from what is most deeply human. Our age of excarnation is marked by increasing individualism, and the loss of deep and meaningful relations: to one another, but even more dramatically to the earth, and especially to earth's wild places. As persons we are irreducibly unique and essentially relational; consumptive individualism and excessive objectivism have exaggerated the former while eroding the latter.

Anacarnation presents an important response to this predicament: a return to the joys of embodied life, purged of sentimentalism, without rejecting the technological advances that authentically enhance what is best in and around us.

Looking on as an outsider at the modern alliance of state power and techno-scientific progress, Nietzsche asks in his third *Untimely Meditation*: "Now, how does the philosopher view the culture of our time?" (Nietzsche 1997, 148). The philosopher steps back, and questions: "When he thinks of the haste and hurry now universal, of the increasing velocity of life, of the cessation of all contemplative-ness and simplicity, he almost thinks that what he is seeing are the symptoms of a total extermination and uprooting of culture" (ibid.). The waters of religion are receding, he goes on to note, and the nations are preparing to destroy each another. Scientific inquiry is pursued without restraint and at any cost, its findings under-mining firmly held beliefs. Educated persons are swept along in the currents of "a hugely contemptible money economy" (ibid.). The greed of the money-makers and the greed of the state have co-opted the counterbalancing force of culture to their own ends, so that now education is geared on the one hand toward "making money as easily as possible" and on the other hand toward the state's "advantage in its competition with other states" (ibid., 164 and 165).

"Nowadays the crudest and most evil forces, the egoism of the money-makers and the military despots, hold sway over almost everything on earth" (Nietzsche 1997, 150). While capitalist consumerism exalts the individual's private wants and desires above the common good of the collective, the greed of the state sacrifices the true flourishing of its members, as part of a larger body, to the competitive ad-vantage of the state-ego. Today we are seeing a synthesis of the twentieth-century extremes of capitalist individualism and communist or totalitarian collectivism in the modern, military-driven nation-state: fueled by the individualized self-interest that drives its consumer economy, and steered by the egoism of belligerent pilots. There is no unifying, substantive vision of *the good* to guide, and all meaningful connections to the larger cosmos have been eroded.[8] The theater in which these theatrics play out, the earth, and the atmospheric conditions that have made human civilization possible, are devastated in the process. This is to our own peril, and to the peril of the species, individuals, and poorer nations that shoulder the unwel-come effects of our combined actions.

"Who is there then, amid these dangers of our era, to guard and champion *humanity*"? (Nietzsche 1997, 150). In our modified sense: who is there to challenge the status quo, and to champion *human responsibility*? It is ours alone to become more gracious guests of the earth, and better hosts to the creatures whose fates now more than ever rest in our collective hands. We must become more attentive to the needs and struggles of the poorest among us, out of sight and out of mind.[9]

We are more connected now than ever, socially, and convenience is literally at our fingertips for the first time in history. We can contact anyone at any time, and within hours nearly anything can be delivered at the doorstep. Yet it would seem that we are more isolated and less satisfied now than ever before in our recorded history. Our callousness begins in childhood: "In terms of the world as we understand it today, embedded in the age of excarnation, capitalism—fueled by competition—and the speed at which we need to move to keep up makes us callous early on. We are less attentive to what is beyond ourselves, less stifled by the pain of the other, less empathetic altogether" (Fitzpatrick 2023, 42). The price of rapid techno-scientific advancement is hyper-individualism, exacerbating the unquestioned byproducts of unrestrained capitalist consumer-culture. The things and systems we have built around ourselves, by and large, have not been constructed with the fullness of our humanity in view. These instead are largely the result of national quests for military expansion, serving a lust for power; and unregulated market forces, serving to meet the unruly demands of unchecked appetites and desires.

In addition to *aestheticization*, as a sensitizing response to the modern problem of anesthetization, we do well to embrace *anacarnation* as a tactful response to the predicament of postmodern or perhaps posthumanist excarnation. Anacarnation is a secondary return to the primary wonders of embodiment, having passed through the night of excarnation (Kearney 2023, 234; 2021, 113ff).[10] It is a recovery of the joys of incarnation without jettisoning the best our technologies have to offer (Treanor and Taylor 2023; cf. also Treanor 2021). It means calling into question the thoughtlessness with which so much of our technology has been constructed, along with the assumption that if we can build something, then we should, and if it can be mass-produced, then every consumer should have one.

Anacarnation would not require simply leaving the city, never to return again, but rather getting in touch with the biotic life already thriving within the city. It would require that *cultivation*, in both built environments and biotic environments, take place more thoughtfully. It would mean developing technologies that assist in securing what we *actually* need, as embodied human persons of a species with a long biological history, arising from deepened understanding of ourselves as contemplative human animals (Kearney 2021, 129ff; McGrath 2019). We need, therefore, to consult the wisdom of those most in touch with the earth, its animals, and one another, counterbalancing our excess with the wisdom of indigenous, and also non-Western traditions. We need massive demands for change, and justice, on the part of holistically educated citizens. These conditions would require, in turn, supporting education that touches to the heart of who and what we are, as persons, by addressing both the interpersonal *and* cosmological dimensions of our personhood.

Anonymization

Anonymity refers to a state in which lack of personal identification predominates. To anonymize, in our sense, is to render nameless. It is to speak of abstract universals like species or ecosystems without engaging in personal, embodied relations to actually existing individuals (cf. especially the Introduction in Dufourcq 2022). It corresponds to anesthetization, or scientific knowing, and to objectification in the absence of appreciation for the non-instrumental value of the being of other creatures, and in the absence of humbled human acknowledgment of our unique responsibility to care. It coincides with alienation, or the perverse coercion of culture by the state-led alliance of consumptive greed and the lust for power. Anonymization is actively resisted by the counterbalancing force of *attestation*.

The objectivistic gaze, a necessary component of scientific knowing, has been especially helpful, for instance, in the objective study of the body in medicine. But when technological advancement and the scientific quest for knowledge were finally wedded, in the early modern period, the hubristic project of becoming masters and possessors of nature soon became powered by a combustion engine. Industrialization, exploitation of expendable workers and nonhumans, and large-scale pollution became ideologically unacknowledged elements of that project.

Previously scientific investigation was reserved for those with the time for leisurely investigation, when the whole endeavor was known as natural philosophy. Technology was the domain of those dedicated to daily labor, typically by hand, where innovations served to facilitate their work. But in early modernity, in theory, and later in the modern period, in practice, technological advancement gradually allied itself with the new scientific enterprise:

> [I]t was not until about four generations ago that Western Europe and North America arranged a marriage between science and technology, a union of the theoretical and the empirical approaches to our natural environment. The emergence in widespread practice of the Baconian creed that scientific knowledge means technological power over nature can scarcely be dated before about 1850, save in the chemical industries, where it is anticipated in the 18th century. Its acceptance as a normal pattern of action may mark the greatest event in human history since the invention of agriculture.
>
> (White Jr. 1967, 1203)

We can see the foundations for this wedding in Bacon's early modern scientific creed, to be sure, but of greater consequence for our purposes is the Cartesian marriage of *philosophical* with *scientific* knowing; namely Descartes' metaphysical physics, to cite the title of a relevant work (Garber 1992). Rather than counterbalancing the objectivistic pursuit of knowledge with grounded philosophical knowing, which keeps in close moral contact with both the love of value and being and the necessity of humbling the self or ego, Descartes subordinates—according to the definitions offered above—philosophical to scientific knowing.

All that is not *mind* is relegated to the status of *extension*, as that which can be objectified and measured. Extended things can then be manipulated, controlled, and dominated by the human will. Nonhuman animals bear the brunt of this theoretical shift. Like human bodies apart from the mind, they are now thought to operate mechanistically; so they can be cut open while still alive, tortured in the name of scientific knowing, and eventually industrially raised, in mass quantities, and consumed daily without a second thought. The ecofeminist insights of Carol J. Adams are especially relevant here: the masculinist epistemic project of analysis through objectification, absent an attitude of care, leads to the inhumane exploitation of both women and animals (Adams 1990).[11]

In our age the philosopher must be called to think carefully about the unquestioned values and overweening hubris behind so many of our inventions, expansions, and acquisitions: to question the presumptions of the Holocene and to push forward toward a new Symbiocene epoch (Kearney 2021, 111). This would require seeking out the advice of those for whom "the whole spiritual person" has come to "love absolute value and being" and for whom "the natural self and ego" has been radically humbled (Scheler 2010, 95).[12] I add that these together should lead toward an attitude of reverence; toward recognition of, and *attestation to*, the *haecceitas* or thisness of each living creature.[13] This would entail both becoming animal *and* becoming more fully human at one and the same time. It would mean first attending to, then attesting to—especially through aesthetic media, in preparation for actual contact—the subjecthood of those who have been anonymized in our age, not least of all our depersonalized and exploited animal kin.[14]

Conclusion

At the heart of the climate emergency is a loss of awareness of the cosmological dimension of our existence as persons. Driven by insatiable desire, and condemned to construct without any real vision of ecological harmony to guide our development, for five centuries the human animal has been exalting itself to its own destruction. With our material constructions we first isolated ourselves from wildness, danger, and surprise; with our immaterial constructions we have further isolated ourselves from each other, from both human and nonhuman animality, and from the earth that hosts our pursuits.

There have been, of course, unquestionable gains with each techno-scientific advance. But in the age of excarnation we have alienated ourselves from the meaning-saturated world that exists beyond our objectivistic ways of knowing, and which subsists beneath or behind our societal constructions. Only once *we* find healing, and accept our unique, creaturely responsibility, can we offer healing to the animals whose general wellbeing is now more than ever in our hands, to the plants and the ecosystems that sustain our relations to all living things, and to one another, as members of one human family.

Ecological degradation is a personal problem, rooted in human desires and (lack of) self-understanding. It demands a personal response. Becoming faithful to the earth would require active cultivation of the remedies we have discussed, namely *aestheticization*, *anacarnation*, and *attestation*. We diminish our own dignity as persons when we instrumentalize the lives of nonhuman creatures, and when we treat with callous indifference the earth that hosts us.[15] We become more fully human when we acknowledge that we are simultaneously guests of earth and hosts to all of the living: when we accept our unique vocation first to humble ourselves, and our egoism, and then to cultivate, care for, and protect earth's diverse creatures and wild places.

Notes

1 "A *technē* which would set humans free from the bondage of drudgery, to be the stewards rather than the desperate despoilers of nature, should surely not be despised" (Kohák 1987, x). In *Thinking Nature*, Sean McGrath has argued persuasively that we must strive to cultivate *contemplative critique, contemplative control, and contemplative calculation* (McGrath 2019, 11; 134–135).

2 See especially Kearney 2023, 238ff; and also the "sacramental aesthetic" at Kearney 2011, 133; cf. Bradley 2023, 96f. In fact, one should consider the sections of the present chapter against the background of Bradley's insightful analysis of the metaphor of "depth" in Kearney's work, in relation to distinctive but overlapping periods of his thought: *moral imagination, carnal hermeneutics*, and a *theopoetics of creation* (Bradley 2023). The three sections to follow ("Anesthetization," "Alienation," and "Attestation") might very well correspond to these three themes in Kearney's thought. I have been deeply influenced by Kearney's thinking, and Ricoeur's, both directly and, in a much larger part, indirectly—especially through close friends and mentors like Treanor and Bradley. The last theme, a theopoetics of creation, is being developed currently, including in the present book project, and especially in dialogue with recent work, cited below, by Treanor, Bradley, Fitzpatrick, and Gschwandtner (Gschwandtner 2023).

3 My thanks to Jeffrey Bloechl for directing me to the passages that follow, from Scheler's text.

4 Bradley concludes: "Unmoored instrumental reason sooner or later always becomes a terror, and we seem even less well-positioned now than thirty years ago to escape instrumentalism. The great and enduring promise of environmental philosophy lies in continually reminding us that the more-than-human world has more than instrumental value—and that this is nowhere clearer than when its existence is threatened" (Bradley 2019, 211).

5 This counterbalancing will require, at least in part, our getting in touch with bodily ways of knowing. Cf. Treanor: "My suggestion is that if we hope to get closer to empathy with our non-human kin, we must bracket or suspend, as much as possible, our all-too-human linguistic engagement with the world" (Treanor 2023, 21).

6 Heidegger included at least one of these two—although Eckhart's original use almost certainly contains reference to both—within the scope of *Gelassenheit*, or letting be, as a response to the techno-scientific mastery: "Heidegger adopted the term to describe twisting free from thinking-as-willing and from the technological worldview that defines the modern epoch, reducing all beings to 'standing-reserve' (*Bestand*), a mere well of resources from which to draw for human wants. He claims in his later works that this voluntarism, reaching an apotheosis of sorts in Nietzsche, is a driving force in the descent of the Western tradition into nihilism" (Treanor 2021, 87). On those points one should consult Bryan Bannon's aptly titled *From Mystery to Mastery*, on Heidegger and Merleau-Ponty (Bannon 2014). In what follows I cite approvingly the early Nietzsche's criticisms of the excesses of his time, in his *Untimely Meditations*. For more critical engagement with Nietzsche's mature thought, in dialogue with Treanor's *Melancholic Joy*, see especially Rogers 2021b.

7 For more on Scheler in relation to these topics, see especially McCune 2014. Kearney makes an aesthetic connection to *Attestation*, to be discussed further in the final section, below, in citing Richard Rohr: "Until we can experience each thing in its specific 'thisness' as artists so often do, we will not easily experience the joy and freedom of divine presence…" (Kearney 2023, 253n23).

8 Nietzsche's prophecy is coming true at both levels: "The revolution is absolutely unavoidable, and it will be the atomistic revolution: but what are the smallest indivisible basic constituents of human society?" (Nietzsche 1997, 150).

9 If haste and hurry, thoughtlessness in the pursuit of excess, exaggerated trust in scientific knowing, the alliance of techno-scientific innovation and military funding, and negligent avarice were all clear symptoms of cultural sickness in Nietzsche's day, what might we, a century and a half later, say in response to the excesses of our own time? To our sixty-hour workweeks and worship of the celebrities who entertain us, to Instagram influencers and snapshot projections of happiness, to Tiktok and Snapchat addictions, to billboard seductions and reality television? To a McDonalds on every street corner, and the billionaire competition to fund private space exploration for the ultra-wealthy? To nuclear proliferation and ultra-polarization, and to the unprincipled and unregulated race to develop the latest AI technology?

10 "In various wisdom traditions, anatheism expresses itself as panentheism—God in all things but not God as all things (…). Such invocations are consonant with the panentheism of Celtic mystics like Eriugena and Pelagius who wrote that God is present in all that has life, meaning not just fellow human but other-than-human life forms that surround us" (Kearney 2023, 253n23).

11 As Fitzpatrick puts it, citing the early modern philosopher Jacques Rousseau and the Harvard psychologist Melanie Joy, for most of us the default attitude toward animals and toward one another "is that we prefer to do no harm." She continues, "But, as Rousseau argues, a certain form of thinking—specifically, competitive calculation and consumption—can desensitize us from our carnal revulsion to suffering, and, as a result, lead us to disregard our own vulnerability and the vulnerability of others" (Fitzpatrick 2023, 41).

12 "Loving, however, is not a preference or a rejection of value, but a movement toward something as bearer of value, a movement that first allows that value to flash forth in and of itself…" (Steinbock 2021, 61).

13 Kearney cites a lesson taught by Breton, with reference to Hopkins: "Each person, Breton taught me, is everyone. The particular is the universal. The concrete is the cosmic. The infinitesimal is the infinite. Epiphanies are ordinary, everyday things. God is a god of little things—the last and least. The strangeness of every stranger" (Kearney 2018, 41). Without taking anything away from the privileged position occupied by human faces, as sites for theophany, with Fitzpatrick and Treanor we can talk about extending such insights to nonhuman animals (cf., in addition, Mizzoni 2008 and 2004).

14 Cf. Treanor 2023, 22; and Kearney and Fitzpatrick 2021, 55.

15 "It is, as [Pope] Francis puts it, 'an act of love that expresses our dignity.' Thus, by acting for the sake of the other that is nature, we further develop ourselves in terms of our human dignity (or species-being in Marxian terms)" (Weidel 2019, 35–36). Cf. also Jeannot: "…Only a humanism that recognizes the full reality of the person, identifies the 'person as [the] key to reality,' and takes 'personal living' as philosophical 'starting point,' can be a fully adequate, integral humanism—and thus provide the philosophical basis for an adequate response to our environmental crisis" (Jeannot 2019, 171).

Bibliography

Adams, Carol J. 1990. *The Sexual Politics of Meat*. New York: Bloomsbury.

Bannon, Bryan E. 2014. *From Mystery to Mastery: A Phenomenological Foundation for an Environmental Ethic*. Athens: Ohio University Press.

Bradley, Daniel O'Dea. 2019. "The Ethics of Sustainability, Instrumental Reason, and the Goodness of Nature: From the Abstractions of Despair Back to the Things Themselves." In *Sustainability in the Anthropocene: Philosophical Essays on Renewable Technologies*, ed. Roísín Lally, 193–214. Lanham: Lexington.

Bradley, Daniel O'Dea. 2023. "Deep Calls to Deep." In *Anacarnation and Returning to the Lived Body with Richard Kearney*, ed. Brian Treanor and James L. Taylor, 86–106. New York: Routledge.

Dufourcq, Annabelle. 2022. *The Imaginary of Animals*. New York: Routledge.

Evernden, Neil. 1999. *The Natural Alien: Humankind and Environment*. Toronto: University of Toronto Press.

Fitzpatrick, Melissa. 2023. "Sensing the Call of Other Animals: Carnal Hermeneutics, and the Ethico-Moral Imagination." In *Anacarnation and Returning to the Lived Body with Richard Kearney*, ed. Brian Treanor and James L. Taylor, 32–48. New York: Routledge.

Garber, Dan. 1992. *Descartes' Metaphysical Physics*. Chicago: University of Chicago Press.

Gschwandtner, Christina M. 2023. "The Embodied Human Being in Touch with the World: Richard Kearney, and Hedwig Conrad-Martius in Conversation." In *Anacarnation and*

Returning to the Lived Body with Richard Kearney, ed. Brian Treanor and James L. Taylor, 49–65. New York: Routledge.

Jeannot, Thomas M. 2019. "Beyond Naturalism: A Personalist Integral Humanism." In *Sustainability in the Anthropocene: Philosophical Essays on Renewable Technologies*, ed. Roísín Lally, 171–192. Lanham: Lexington.

Kearney, Richard. 2011. *Anatheism: Returning to God After God*. New York: Columbia University Press.

Kearney, Richard. 2018. "Where I Speak From: A Short Intellectual Autobiography." In *Debating Otherness with Richard Kearney: Perspectives from South Africa*, 31–62. Cape Town: AOSIS.

Kearney, Richard. 2021. *Touch: Recovering Our Most Vital Sense*. New York: Columbia University Press.

Kearney, Richard. 2023. "Anacarnation: Recovering Embodied Life." In *Anacarnation and Returning to the Lived Body with Richard Kearney*, ed. Brian Treanor and James L. Taylor, 233–256. New York: Routledge.

Kearney, Richard and Melissa Fitzpatrick. 2021. *Radical Hospitality: From Thought to Action*. New York: Fordham University Press.

Kohák, Erazim. 1987. *The Embers and the Stars: A Moral Inquiry into the Moral Sense of Nature*. Chicago: University of Chicago Press.

McCune, Timothy J. 2014. "The Solidarity of Life: Max Scheler on Modernity and Harmony with Nature." *Ethics and the Environment* 19:1, 49–71.

McGrath, Sean. 2019. *Thinking Nature: An Essay in Negative Ecology*. Edinburgh: Edinburgh University Press.

Merleau-Ponty, Maurice. 2012. *Phenomenology of Perception*. Translated by Donald Landes. New York: Routledge.

Mizzoni, John. 2004. "St. Francis, Paul Taylor, and Franciscan Biocentrism." *Environmental Ethics* 26:1, 41–56.

Mizzoni, John. 2008. "Biocentrism and the Franciscan Tradition." *Ethics and the Environment* 13:1, 121–134.

Nietzsche, Friedrich. 1997. *Untimely Meditations*. Cambridge: Cambridge University Press.

Nietzsche, Friedrich. 2005. *Thus Spoke Zarathustra*. Oxford: Oxford University Press.

Rogers, Chandler D. 2021a. "Being Consistently Biocentric: On the (Im)possibility of Spinozist Animal Ethics." *Journal for Critical Animal Studies* 18:1, 52–72.

Rogers, Chandler D. 2021b. "After Dark: Nullifying Nihilism." *Journal of the Pacific Association for the Continental Tradition* 4, 184–190.

Rogers, Chandler D. 2023. "Reverence for Life and Ecological Conversion." *Worldviews: Global Religions, Culture, and Ecology* (forthcoming).

Scheler, Max. 2010. *On the Eternal in Man*. New York: Routledge.

Steinbock, Anthony J. 2021. *Knowing by Heart: Loving as Participation and Critique*. Evanston: Northwestern University Press.

Treanor, Brian. 2021. *Melancholic Joy: On Life Worth Living*. New York: Bloomsbury.

Treanor, Brian. 2023. "Thinking Like a Jaguar: Carnal Hermeneutics, Touch, and the Limits of Language." In *Anacarnation and Returning to the Lived Body with Richard Kearney*, ed. Brian Treanor and James L. Taylor, 13–31. New York: Routledge.

Treanor, Brian and James L. Taylor. 2023. *Anacarnation and Returning to the Lived Body with Richard Kearney*. New York: Routledge.

Weidel, Timothy A. 2019. "*Laudato si'*, Marx, and a Human Motivation for Addressing Climate Change." *Environmental Ethics* 41:1, 17–36.

White, Lynn Jr. 1967. "The Historical Roots of Our Ecologic Crisis." *Science* 155, 1203–1207.

Part IV

Philosophies of the Earth

Chapter 9

Faithful to the Earth: Hölderlin, Nietzsche, Rilke, and Heidegger

Joseph S. O'Leary

The physical threats to our planet inspire terror and rage. But we are also losing the earth, or failing to care for it, in a subtler way, namely through the disappearance of Nature as a mystical, poetic presence. English Literature is no longer a bulwark against such forgetting. Scientific ecology, ironically, undercuts poetic vision, discards it as naive. Nature writing has been a booming industry among literary critics in recent decades, but one somehow felt that the poets one most enjoyed could not advance under such a virtuous heading. Now I would like to call them up, not for literary purposes only but more as a *cri du coeur*.

Exiled from the earth in the world's most populous city, and meeting Nature only in the shrubs on the nearby well-tended path running all the way from Ogikubo to Takadanobaba between the backs of houses over a buried stream, or in the immaculate winter skies, I treasure especially a few wayside Shinto shrines, ancient and unnoticed. This religion persists quietly, despite the dry treatment of it by the scholars,[1] and despite Japan's frenetic construction industry,[2] and it holds out a plateau for an interreligious encounter around the sacred presence of the earth and for mystical deepening of ecological discourse. Indeed it is ripe for promotion to the status of a universal religion on this basis,[3] and the earth-poets would find a home in that religion.

Wordsworth, whose witness to the mystical presence of the earth is uncluttered by complex mythology—unlike Hölderlin's—and who did not decline into madness but into gentle elegy in the line of Gray, is travestied in a literary criticism more interested in politics, gender, and semantics than in the irreducible sublime event his most inspired verse attests.[4] He tried to Christianize his pantheistic vision, but at the cost of diluting it. G.M. Hopkins and Paul Claudel, in contrast, could be robustly chthonic and biblical at the same time: *"Comme toute la création est avec Dieu dans un mystère profond."*[5]

Turning to the Germanic realm, we find a persistent conflict between the chthonic sublime and orthodox Christianity in Friedrich Hölderlin (1770–1843), Friedrich Nietzsche (1844–1900), Rainer Maria Rilke (1875–1926), and Martin Heidegger (1889–1976), all of whom must concoct new redemptive philosophies against the somber background of the flight of the gods, the death of God, or the technological depletion of being. Behind them lies Goethe's joyful rapport

DOI: 10.4324/9781003456940-13

with Nature, which they cannot retrieve; nor can we. There is an inner bond between the German landscape and the German language: they share a singular quality of *presence*, which English translations of the poets only faintly reflect, and against which Anglo-American monolingualism is a sure defense.[6] When Eichendorff wrote the lines: "*Da steht im Wald geschrieben/Ein stilles, ernstes Wort*" (There is written in the wood a silent, serious word), his verse itself fuses with the silent peal of nature. No doubt every language, each in its own unique way, can attain moments of such magical accord with the earth's enveloping and enigmatic presence.

These poets do not need to be brought safely back within the ambit of Christian ecology as articulated by Pope Francis in *Laudato si'* (2015), or forced to be made useful in ecological propaganda of any kind. But care for the earth[7] can be anchored and enriched by contemplative attunement to their voices, however obscure or "pagan" their message. Recitation of the earth-poets should be prominent in our liturgies.

In "*Wie wenn am Feiertage...*" (1800), the first of his "Hymns," modeled on Pindar, which follow on the "Elegies" (such as the breath-taking "*Brod und Wein*" and "*Heimkunft*"), Hölderlin embarks on a more ambitiously symbolic and myth-making embrace of nature, with a more intensively reflected articulation of his historical and eschatological mission as poet.[8]

Jetzt aber tagts! Ich harrt und sah es kommen,
Und was ich sah, das Heilige sei mein Wort.
Denn sie, sie selbst, die älter denn die Zeiten
Und über die Göttern des Abends und Orients ist,
Die Natur ist jetzt mit Waffenklang erwacht.[9]

(But now dawn is breaking! I waited and saw it come, and what I saw, the Holy be my word. For she, she herself, who is older than the times and is above the gods of East and West, Nature has now awakened with the clangor of arms.)

At all times Hölderlin could speak out of a deep contemplative attunement to the earth, and if one reads him only for that he is still a supreme figure among modern poets. But his ambition stretched beyond this, carrying German philhellenism to an extreme pitch. His engagement with Nature became a revolutionary battle to save the earth, making Schiller's dream of universal fraternity come true and inviting the long-lost gods of Greece to return. Or rather, as Emil Alexandrov reminds me, those gods have become ciphers of Nature, *phusis*, or Being, as Heidegger will interpret them, and when Empedocles leaps into Etna to be reunited with them, he is a martyr of the mystical Earth rather than of its former heavenly visitants.[10] It is disillusioning to discover that Hölderlin's Sophocles translations, over which everyone laughed at the time, contain egregious errors in every 4–5 lines.[11] It is not surprising that Heidegger was full of dread when he visited Greece, fearing that the Greece Hölderlin and he himself had concocted would be exposed as a fatuous delusion.[12]

Hölderlin remained in touch with Nature to the end, penning childlike celebrations of the seasons for the visitors to that sunny tower over the Neckar where he pottered about for more than thirty years of mental obnubilation:

Ihr lieblichen Bilder im Tale.
Zum Beispiel Gärten und Baum,
Und dann der Steg, der schmale,
Der Bach zu sehen kaum.

(Ye lovely images in the valley, for instance gardens and tree, and then the narrow footbridge, the brook scarce to be seen.)[13]

Note the touching echo of Goethe's *Nähe des Geliebten* (well known through its Schubert and Schumann settings): "*In tiefer Nacht, wenn auf dem schmalen Stege/Der Wandrer bebt*" (In deep night, when on the narrow footbridge, the wanderer trembles).

At seventeen years old, Nietzsche wrote a glowing essay on Hölderlin, part of his own awakening, but later he rejected Hölderlin's "Platonism" which embraced the value of Truth rather than Will-to-Power.[14] He inherited the poet's passion for Alpine landscapes, and his Zarathustra also descends to the valleys, woods, and plains where common mortals dwell.

The overman is the meaning of the earth. Let your will say: *Be* the overman the meaning of the earth. I beseech you, my brothers, *remain faithful to the earth,* and do not believe those who speak to you of hopes beyond the earth.[15]

Nietzsche today is probably preferred as a waspish culture critic rather than as an Alpine visionary, and many will find the over-wrought rhetoric of *Zarathustra* wearisome. Yet it is in his love of the earth that he strikes his highest notes, and rejoins his *bête noire* Wagner, whose genius dominated and obsessed him throughout his career. For what is most attractive about Wagner, too, is the radiant presence of the earth. No music is more porous to natural sublimity: the sea in *The Flying Dutchman* and *Tristan*, the Rhine, the spring night, the forest, the rocky heights as they leap into life in the *Ring*—though again this is an aspect one hears little about now and that *Regie-Theater* performances cannot convey.

Rilke was only one degree of separation from Nietzsche, as the inheritor of his girlfriend Lou Andreas-Salome. He felt that the earth was being stolen away, and that technology, which he hated, had robbed fields and woods, plants and "things" more generally, of their authentic presence. The *Duino Elegies* find resolution to their problematic of human existence in the Ninth Elegy's climactic celebration of the earth in the here and now,[16] and promise to bring the earth into its own in the inner space, the *Weltinnenraum* of human consciousness.

Are we, perhaps, here just for saying: House,
Bridge, Fountain, Gate, Jug, Olive tree, Window…?

Earth, isn't this what you want: an invisible
Re-arising in us? Is it not your dream
To be one day invisible? Earth! invisible!
What is your urgent command, if not transformation?[17]

The listed things are familiar items, like that *Steg* in Goethe and Hölderlin, or "the little hill of Hermon" in Psalm 42:6 or "some tree on a slope, to be looked at day after day" in the First Elegy, which become quilting points linking the subject to the natural world. Is sentimental projection, the pathetic fallacy (or what D.H. Lawrence denounced in Wordsworth as "making flowers speak") at work here? In the poetic metamorphosis does Rilke lose the real presence of the earth? Romano Guardini's summary kindles that suspicion: in poetic saying "the thing acquires a density and a sense of being that it could never attain from itself. Once such a word is spoken the thing 'thinks': 'now at last I really am.' Not in the form of the first, immediate being, but of the second, of 'inwardness (*Innigkeit*).' The first is mere being-at-hand (*Vorhandenheit*)."[18] These are questions for reflection, not formulas of dismissiveness.

Heidegger's lecture, "*Wozu Dichter?*"[19] on the twentieth anniversary of Rilke's death in 1946, suggests that the poet is trapped in a Nietzschean metaphysics of subjectivity. Taking the title from Hölderllin's "Wherefore poets in a time of need?" ("*Brod und Wein*") it consigns Rilke to his due place, a lesser one, in Hölderlin's tradition. After the debacle of his Rectorship Heidegger had wrestled with his German heritage in his copious lectures on Hölderlin and Nietzsche from 1934 to 1943.[20] He presents Nietzsche as the end-point of the alienation of Western metaphysics from Being and Hölderlin as the thinker who puts us back in touch with the Greek beginning. Against Rilke, and emboldened by Hölderlin, Heidegger interprets the earth as the presence of Being, not to be absorbed into the subjective sphere of consciousness or imagination. The Earth images Being in its withdrawal.[21] "Nature likes to hide itself" (Heraclitus). Commenting in 1939 on the lines from "*Wie wenn am Feiertage...*" quoted above, Heidegger writes:

Nature is more timely (*zeitiger*) than "the times," since as the wonderfully all-present it has already previously granted to everything actual (*wirklichem*) that Clearing (*Lichtung*) into whose Open (*Offenes*) whatever can be called actual is first able to appear.[22]

The extraordinary fusion between poet and philosopher, notably at the level of vocabulary, lends Hölderlin and his Greece an extraordinary authority in the world of philosophical thinking, which can be justified only if the mystique of the Earth that they share is really grounded in a perception of fundamental reality, of Being.

Contemplation is key here. My late lamented friend Hubert Durt had an uncle who was steeped in literature. Once in a hotel in Belgium he noted the fussiness of another guest, complaining bitterly that his artichokes had not been shaved as he demanded. Curious, he asked the waiter who the guest was. "A complete lunatic," was the reply, "he spends hours gazing out his window at the stream." The guest was Marcel Proust. Yet when Jean Beaufret tried to get Heidegger interested in Proust, he replied: "Balzac is closer to the Greeks." Just as students of mysticism such as Adolphe Tanqueray (1854–1932) would grade degrees of contemplative depth, so Heidegger had his demanding canon of authenticity for the saying and thinking of being, and many great names did not make the grade. He caused offense by referring to Goethe's "sing-song" and, as Beaufret also informed me, he much preferred the Brandenburg Concertos to Wagner ("the Nazi music"). To be faithful to the earth is an ascetic discipline and an education in taste, even demanding costly renunciations.

Samuel Beckett in April 1951 commented on Heidegger's elucidation of "*Wie wenn am Feiertage...*" and an essay by Maurice Blanchot: "*À mon avis ce n'est pas la peine de sauver Blanchot. Des extraits de Heidegger, mal traduits par moi d'une traduction française déjà fort embarrassée, ne pouvant y apporter le moindre éclaircissement, au contraire... Que c'est cotonneux tout ça et que le pauvre Hölderlin est mal tombé.*"[23] Is Heidegger as "woolly" as the French translation (absurdly translated in turn into English by Beckett) made him seem? Though published in the *Hölderlin-Jahrbuch*, Heidegger has not had a deep impact on the erudite guild of Hölderlin scholars. They are unable to learn from his sounding of the rapport between thought and poetry, which problematizes categories the philologists simply take for granted.[24] As always, Heidegger reads with his own concern with Being in mind, in this case entangled with a myth of Germany's privileged role in the history of Being.[25] It is too easily presumed that Heidegger is philologically arbitrary and incompetent, but Hermann Mörchen has shown that Theodor Adorno's dismissive treatment of Heidegger on Hölderlin, in a famous essay, is based on careless reading.[26] True, since Heidegger's interpretation of Hölderlin springs from "a necessity of thought"[27] "violence is sometimes done to the text."[28]

Reading the poets "in a time of need" for the earth today we must inevitably solicit them into an orbit they did not suspect, while at the same time we let them draw us into theirs. Taken up in a revisionist style, the accord between Heidegger's thinking of Being and Hölderlin's naming of the Holy, both of which are rooted in awareness of the presence of the earth, may provide a deep anchorage for ecological awareness, saving it from frenetic anxiety. The earth speaks to us through their voices, indeed pleads with us to remember its precious, enfolding presence. I recall them here as only one, rather hoary strand of earth-awareness, to be connected with that of countless tribes throughout the planet, such as Sean McDonagh's beloved Tboli people in the Philippines, Robin Kimmerer's indigenous Potawatomi nation or Pope Francis's devotees of Pachamama (Mother Earth). All have their sophisticated ways of saying, with Rilke, "Earth, you darling, I will!" Shared care for the earth can unite people across the widest range of cultures and may gird the planet with a protective vest of poetry, as essential to its survival as the ozone layer. Ecological awareness need not be a diet of gruel, but can reopen treasures of responsive imagination, making our threatened planet a more joyful, colorful, and livable place.

Notes

1   See for instance Michael Pye, ed., *Exploring Shinto* (Sheffield: Equinox, 2020), and my review in *Japanese Journal of Religious Studies* 48 (2021): 198–201.
2   See Alex Kerr, *Dogs and Demons: Tales from the Dark Side of Japan* (New York: Hill & Wang, 2002); James W. Heisig, "Make-Believe Nature," *Japan Christian Review* 59 (1993): 103–111.
3   See the Green Shinto website, and Katō Taishi, "The Universality of Shintō, Clarified by Some Buddhist Concepts," *Japan Mission Journal* 72 (2018): 10–18; "The Universality of Shintō as Seen in the Enthronement Ceremonies," *Japan Mission Journal* 73 (2019): 75–78.
4   See J.S. O'Leary, "A Mystical Utterance in Context: 'Tintern Abbey.'" *Sophia English Studies* 38 (2013): 19–58.
5   *L'annonce faite à Marie*, in Paul Claudel, *Théâtre*, II (Paris: Gallimard, 1956), 136.
6   See Sean McDonagh, *To Care for the Earth: A Call to a New Theology* (London: Chapman, 1986). According to his mentor Matthew Fox, Sean played the foremost role in the composition of *Laudato si'*. The Pope's call to "ecological conversion" in that Encyclical is laced with poetic consciousness of a personalized earth, which may owe something to Hölderlin, his favourite poet, and to Rilke, both of whom were studied in brilliant monographs by Romano Guardini (1885–1968), the topic of Pope Francis's postgraduate study.
7   See McDonagh, *To Care for the Earth*.
8   See Bernhard Böschenstein, *"Frucht des Gewitters": Hölderlins Dionysos als Gott der Revolution* (Frankfurt: Insel, 1989), 114–136.
9   Friedrich Hölderlin, *Gedichte*, ed. Jochen Schmidt (Frankfurt: Deutscher Klassiker Verlag, 1992), 239.
10 See Friedrich Hölderlin, *The Death of Empedocles*, ed. and trans. David Farrell Krell (Albany, NY: State University of New York Press, 2008).
11 See Wolfgang Schadewaldt, "Hölderlins Übersetzung des Sophokles," in Jochen Schmidt, ed., *Über Hölderlin* (Frankfurt: Insel, 1970), 237–293, 246. But Schadewaldt calls them "creative errors" (247).
12 See Maria Villela-Petit, *Questioning Greece with Heidegger and Simone Weil* (Nagoya: Chisokudō Publications, 2023), 201, quoting Heidegger, Gesamtausgabe 75.217.
13 *Sämtliche Werke und Briefe*, 458. Samuel Beckett liked these lines; see Mark Nixon, "Beckett and Germany in the 1930s: The Development of a Poetics," in S.E. Gontarski, ed. *A Companion to Samuel Beckett* (Oxford: Wiley-Blackwell, 2010), 130–142, 139.
14 See Henning Bothe, *"Ein Zeichen sind wir, deutungslos": Die Rezeption Hölderlins von ihren Anfängen bis zu Stefan George* (Stuttgart: Metzler, 1992), 50–54.
15 Friedrich Nietzsche, *Also sprach Zarathustra: Ein Buch für alle und keinen*, Kritische Studienausgabe 4 (Berlin: de Gruyter, 1988), 15.
16 See Käte Hamburger, *Rilke: Eine Einführung* (Stuttgart: Klett, 1976), 149–158
17 *The Duino Elegies*, trans. J.B. Leishman and Stephen Spender (New York: Random House, 1992).
18 Romano Guardini, *Rainer Maria Rilkes Deutung des Daseins: Eine Interpretation der Duino Elegien* (Munich: Kösel, 1953), 344.
19 In *Holzwege*, Gesamtausgabe 5 (Frankfurt: Klostermann, 1975ff), 269–320.
20 See the lecture courses in Gesamtausgabe, Voll. 39, 52, and 53, and the texts collected in Vol. 75. Hölderlin is liable to crop up at any moment in the later Heidegger, whose vocabulary is largely fashioned on the poet's; Max Kommerell wrote to Heidegger in 1942 warning him that such hermeneutical violence could be "a disaster!?"; quoted, Charles Bambach, "Who is Heidegger's Hölderlin?" *Research in Phenomenology* 47 (2017): 39–59, 44.

21 See the pages on the struggle betweeen *Erde* and *Welt* in the constitution of a work of art in "Der Ursprung des Kunstwerkes," Gesamtausgabe 5.1–74, 28–36.

22 Heidegger, *Erläuterungen zu Hölderlins Dichtung*, Gesamtausgabe 4.59.

23 George Craig, ed., *The Letters of Samuel Beckett 1941–1956* (Cambridge: Cambridge University Press, 2011), 230–231. Blanchot's "La parole 'sacrée' de Hölderlin" appeared in *Critique* 1.7 (Dec. 1946), along with Heidegger's essay. Though collected unaltered in *La part du feu* (Paris: Gallimard, 1949), the admiring Leslie Hill's description of it as "slapdash scholarship" is a charitable oxymoron; see Hill, *Blanchot: Extreme Contemporary* (London and New York: Routledge, 1997), 91. Dieter Henrich, authoritative commentator on German Idealism and Hölderlin, has devoted a book to what might seem a tenuous theme: *Sein oder Nichts: Erkundungen um Samuel Beckett und Hölderlin* (Munich: Beck, 2016).

24 See Rainer Nägele's review of Gesamtausgabe 35 in *The German Quarterly* 55 (1982): 100–104.

25 For critical comment on this, see Villela-Petit, *Questioning Greece*, 180–188, 200–210.

26 Hermann Mörchen, *Adorno und Heidegger: Untersuchung einer philosophischen Kommunikationsverweigerung* (Stuttgart: Klett-Cotta, 1981), 99–109. "Does Adorno's Argus-gaze make him text-blind?" (101).

27 Gesamtausgabe 4.7.

28 Mörchen, 102.

Chapter 10

At the Threshold: Nature's Art of the Possible

Christopher Yates

To speak of hosting the earth is to place before ourselves a certain kind of intervention. It is, following thinkers of "hospitality" like Jacques Derrida, Richard Kearney, and Bernhard Waldenfels (among others), to allow ourselves to be called into question, dispossessed of our comfortable operating social norms, and perhaps open to gifts we did not know we needed.[1] To "host" is not so much to welcome something or someone on our terms, but to welcome what might alter those terms for the better. And when, today, the guest is the earth, hosting is most likely counterintuitive, for such a guest is something our way of life is accustomed to occupying, subduing, and marshaling in the name of instrumentalist needs and goals. Ironically, also, nothing to date has been more scientifically explained than the earth, in which case hosting it anew would mean, paradoxically, permitting it to arrive at the door as an irreducible mystery. What if that mystery met us at the threshold with a certain sort of deconstructive grace? Consider the following illustration as a motivating template.

Early in Terrence Malick's 1998 film, *The Thin Red Line*, a story based on America's Guadalcanal campaign of World War II, a rifle company disembarks from a ship and rushes a beach to form a point of entry into the jungle and hills beyond. The transition pivots on a sharp contrast between the soldiers' adrenalized landing and the unexpected quiet of a vacant Melanesian coast. Arriving with a feverish alertness for a battle that does not materialize, their sharp attentiveness takes a topological turn. Soon they are surrounded by nature, as though seeing its wild and encompassing mystery for the first time. There is the chest-high grass of a meadow, then the sunlight filtering down through a high forest canopy, parrots nimbly perched on bladed flora, and the creak of bamboo growing in tall bunches. A soldier asks in voiceover: "Who are you who live in all these many forms? ... You're death that captures all? ... You too are the source of all that's gonna be born... You're glory, mercy, peace, truth... You give calm a spirit, understanding, courage, the contented heart." The voice, in these reverent interrogatives, is addressing earth but more than earth. It is addressing nature as life, as agent, as *arche*. What Malick has done is draw us with his characters across the threshold from the presumptiveness of human *technē* to the exilic mysteries of nature's *physis*.

DOI: 10.4324/9781003456940-14

In the language of phenomenology (a language he well knew), he has affected a movement from the natural attitude to which we are accustomed, to the immediate givenness of the predeterminate field.

Life is seldom so cinematic, but *Who are you who live in all these many forms?* well captures a topological mode of attention which we must venture today. That it is a question affixed to place likewise matters a great deal, for it is in place where the constitution of hospitality materializes. And, environmentally speaking, as it is precisely where we find the degradation wrought by our industrialized instrumentalism over the earth, place must be the zero point from which we take up a receptive hermeneutic that might allow the way of nature to address the way of *anthropos* in a critical yet recapacitating way. It is in this spirit of hosting the "possible" that "moves in all these many forms" that I will turn to two philosophers with earthward visions that I think ought to become our own. Erazim Kohák, working in the spirit of phenomenological description, will direct us to the *moral sense* at work in nature's domain. Friedrich Schelling, working in the framework of Idealist intuition, will direct us to the *artistic striving* in the same. Their point of convergence, we will see, lies in nature regarded as *physis*, and in the corresponding call upon human agency to honor that norm.

Kohák on Nature as the Realm of Moral Revelation

In his 2000 overview of ecological ethics, *The Green Halo*, Kohák laments the presuppositions and practices that have hardened into our "civilizational strategy" of "increasing material consumption" and its "merciless logic of disregard" for nature (Kohák 2000, 5). Our presumptions of "progress" and "development" have set us on a course in which "an infinite demand finds itself in an inevitable conflict with a very finite world" (6, 57). Prone to see the earth as a body of raw materials for our use, we have obscured its "intrinsic value," effectively denaturalizing nature and alienating ourselves from a partnership with it in the vocation of moral dwelling (53). The concern is earnest for Kohák, and when he contends that the "purpose of ecological thought is not theory but practice" he means not only a change in our way of life but also a reorientation by way of contemplative praxis (155).

His prior work, *The Embers and the Stars* (1984), was born of such an endeavor. Having extracted himself from the bright lights and infrastructure of Boston to live for a time in a cabin retreat in a New Hampshire forest, the book (that is all too neglected today) blends poetically styled prose with an eco-phenomenological method in order to present the contours of what he calls a "generic naturalism" (Kohák 1984). His central objective is to let the vital moral order within nature— "what we have hidden from ourselves" (1984, xi) on account of our world of artifacts, artifice, and doggedly physicalist thinking—show itself afresh and rival what has become "a fundamental discontinuity between humans and the natural world" (5). It is a threshold sort of book, an exercise in passing "beyond the powerline and the paved road" and into a dusk that is "pure between the glowing embers and the distant starts" (xii). Cut from the cloth of Husserl, Heidegger, and Aristotle, he carries out a bracketing of that which has heedlessly bracketed the earth in its primary signification.

But the project is not a reduction back to some romanticized prelapsarian milieu (1984, 24). *Technē* (the field of human skill and know-how in making things and constructing culture), he argues in Heideggerian fashion, "can be an authentically human mode of being in the world" (xii). But *technē* detached from *physis* (nature in the mode of purposive growth, becoming) swells into a "Promethean defiance" that stands against the earth rather than finding in it "the meaningfully ordered web of purpose within which each being... has its appointed task" (9).[2] A *technē* that forgets how the human "belongs integrally within a cosmic order" (10) betrays this task and takes on a shunting conceptual and psychological course when it sees nature as resource rather than *kosmos*, object rather than place, something inert rather than vital. Rather than taking its bearings from nature as "the totality of the given," our consumerism and technologism have created "the illusion of autonomous functioning" (22, 25). As strangers to nature, he seems to say, it is no wonder we have become strangers to ourselves, for we have detached ourselves from the moral sense of nature and thereby "the moral sense of our humanity" (13).

Though Kohák does not employ the language of "hospitality," the way in which his work centers on a suspension of human autonomy aligns well with what such an endeavor intends. The path to generic naturalism starts with an important decision to surrender the terms on which we play host. Facing outward from our accustomed domain, we must let nature address us at the threshold and draw us away from our territorializing agendas. Kohák structures his account of what we encounter according to five headings: *Theoria, Physis, Humanitas* (with emphasis on *Cultus*), *Skepsis*, and *Credo*. The first, *theoria*, arises in the about-face and indicates the reorientation incumbent on our habits of mind. Namely, we learn to identify and lay aside our hidebound tendency to conceive of nature as a mechanistic sum of material stuff that, though we make overtures to its life, we effectively deaden by making it speak and serve on the basis of our constructs and interests. Even in its well-intended aspect, our theorizing about all that is "natural" has the adverse effect of distancing us from our true "experiential grounding" (1984, 23). A reform within the disposition of *theoria*, then, would mark a change of attitude as we peer over the threshold.

The second phenomenon, *physis*, names what that shift would encounter—nature not as a body of resources, but as a mode of *poiēsis*; indeed, not nature as the object of our bright daylight's *Denken* (thinking) but as its own dusk-hued *Dichtkunst* (poetics). Here a bit of context is needed. In its early Greek formulation *physis* (what comes to be the Latin *Natura*) lay across the threshold from the realm of *nomos*—human laws and conventions. *Physis* consisted of the growth of plants and of natural being writ large, the life and movement (*kinesis*) of physical birth and becoming. Such a natural art, as Aristotle already knew, would doubtless fall into a relationship with the art of human making—*technē*—in terms of our objects and actions. That is not to say the two phenomena are originally opposed. *Technē* was vested with "imitating" (as in *mimesis*) the production inherent to *physis*, a relationship Heidegger would conceive in terms of *technē* as a bringing-forth into being (*poiesis*). But, as *technē* amasses more power into its production and state of mind, it settles into a contest (as opposed to collaboration) with nature's native art. So it is that, according to Heidegger (and Kohák agrees) the 20th century saw human culture morph into an all-encompassing *Gestell*, an "enframing" agenda that grew comfortable "challenging" nature to deliver its goods. We came to the threshold enamored of the power to make and resource what we want of nature rather than honor the terms/model set by *physis*.

It is with these points of reference in mind that Kohák invites us to return to the realm of "dusk." Why dusk? First of all because it chastens *technē*, dispossessing it of its production-line vision and its commandeering state of mind. Dusk, says Kohák, is "the time of philosophy" (1984, 33); dimming the lights of visual mastery in our *technē*, it preserves wonder and mystery. More than that, however, it is also a time that attunes us to "gifts" issuing from nature's own life-bringing drive. Night grants us nature "in the integrity of its own being" (74). Even the solitude and pain of night are gifts in that they detach us from the false comforts of "conceptual and technological mastery" and make us "willing to see, to hear, to receive" with renewed wonder (46). In a fascinating point, Kohák says that *physis* also offers a special gift of Word to philosophy's own inherent dependence on metaphor, a hermeneutic countermeasure to what is otherwise a detachment of the word of discourse from the text of the "lived reality it presents" (53–54). He explains: "Metaphoric usage [in philosophy] is appropriate to it because *reality itself is metaphoric*; it is the *sense*, not merely the fact or the theory, of being which constitutes its reality" (55). The power of metaphor, that is, rests in the communication of meaning by evoking it. But what is the word of nature that is given to philosophy in this constitutive way? The answer is the centerpiece of Kohák's generic naturalism: the "sense" offered in nature's "word" contains the gift of moral law. If our words are the basis upon which we mete out our technological agenda and enframe the earth according to consumerist campaigns, it follows that their logic of heedless disregard must be the most precise point of reformation in our way of being. Nature is not simply "there," but is a text having its own "meaningful presence" (69). That presence contains an "order of value," a vital order (71). The order, he continues, arises at the intersection of time and eternity in *physis*, which brings with it something constitutive of our awareness of moral rights and wrongs. All this is a matter of a fittedness in the heart of the order of things, a sort of ligature or Pythagorean proportionality connecting the "truth of the order of nature," the truth of "the place of the human therein," and "the truth of eternity" (85).

Where *physis* grants us a constitutive point of moral reference, *humanitas* signals the opportunity to recover the true sense of culture—or *cultus*—on that basis. Over and against our "bulldozer" mode of existence, *cultus*, as cultivation, is a matter of dwelling in and with the earth on non-utilitarian terms (1984, 90). He explains: "It is as beings capable of seeing our place in nature from a moral point of view that we can cease being marauders and can become dwellers in the earth" (91). That possibility involves us in an ongoing high-stakes task. *Cultus* is a project that must be warranted, and its warrant lies in affirming the way "the moral sense of nature" and "the moral sense of being human" are connected (91). When the two are alienated, it falls on us "to ask what *justifies* our presence on this earth at all, since that presence is no longer merely a fact but also a moral problem" (92, my emphasis). The point is less that *humanitas* must "prove" its worth to nature, more that we must live up to the moral project that nature expresses in the "sheer givenness of being" (96). *Cultus*, in this way, is always under an imperative. But, recalling his formulation of the "gifts" that come with dusk, Kohák finds the imperative held within a larger context of natural grace.

Kohák is also well aware that the problem of human "justification"—for those who dare to take it on—can become overwhelming. Under the heading *Skepsis*, he cautions against the temptation to think that we are subjects ultimately fated to an unjustified and alienated state. Having reduced nature to a moveable piece and detached it from teleological coherence, we may incline to do the same with ourselves. Having grown expert at belittling the earth we are all too poised to belittle ourselves, perhaps even unwittingly. A self-critical *cultus* is called for, but *skepsis* amounts to a failure of imagination and foreclosure on the art of the possible in nature. It becomes a solipsistic *fate-accompli*, roping off any openness to the moral sense of nature and to personhood on those terms. The way around the *skepsis* predicament, for Kohák, is to wager an annealing affirmation, a faith—a *credo*. The willingness to utter "I believe" is how we redirect ourselves toward nature's "incoercible givens" as the "outlines of a reality" that evokes a sacred presence (1984, 182, 186). The Psalmist in Kohák sounds in this *credo*. Some readers might worry that a natural theology is covertly operating under the auspices of a natural philosophy. But Kohák's theism enters his text as a phenomenological experience of God, not a doctrine about God. He might well ask, to be fair, "On what basis can we rightly assign the burden of proof to the sacred rather than the secular?" I think this aspect of his *credo* can, moreover, be taken as (to borrow Richard Kearney's term) an "*ana*theistic" openness (Kearney, 2010), a dialogical consideration *in situ* as opposed to a monological finality in sum. Perhaps the possibility of encountering a *Deus sive natura* ought not be discounted altogether as we cross nature's threshold. Regardless, the broader point of *credo* is to pledge oneself and one's sense-making to the purposive poetry that lives "in all these many forms."

Overall, what Kohák has offered in his meditation is a field of possibilities awaiting us on nature's side of the threshold. Should we have the courage to undergo a *kenosis* of our artifactually obsessed habits of a *technē*-gone-rogue and reposition ourselves before the agency of earth's abiding *physis*, we may enter a space of restoration and justification. At the heart of this space stands the expression of a lawfulness coursing in the arterial network of nature's "sheer givenness" which charges us to better integrate the tissue of *cultus* with the constitutive body of an antecedent *logos*. By turning our *theoria* back to this *physis*, and holding the dismay of *skepsis* at bay, Kohák implies an orientation to earth's own distinct *poiesis*. And yet, he does not go so far as to thematize directly another feature housed in nature's array of "gifts"—what we may call *aesthesis*. The attention to the "metaphoric" come closest insofar as it suggests a creative generation of sense in the order of things, but Kohák does not quite see this phenomenon through to the full scope of nature's artistry. Since so much of the problem within the infrastructure of our own abode's *technē* comes down to a coercive and opportunistic craft, what is needed is an instructive countermeasure. Suppose there were a model for us of a "making" capable of finding "justification" in its allegiance to nature's intrinsic craftsmanship?

Schelling on Nature as the Ground of Aesthetic Creation

To turn to Friedrich Schelling's philosophy may seem like a bit of anachronistic "reaching" since the Schelling we know, if we know him at all, tends to be summarily confined to dusty anthologies of "German Idealism" and its "system"-driven house of cards. It is true that Schelling's post-Kantian attempt to reconcile reason and reality—more specifically, intelligence and nature—by uncovering the organizing "system" of all things, may not have been what, by today's press-catalog designations, we would call an environmental philosophy. But neither was it an exercise in aired speculation that would seek to conscript the object world of earth in the ascent of some absolute ground for knowledge via the relationship between human self-consciousness and trans-historical spirit. In a move that was remarkable for its day, Schelling turned toward nature as the herald of a "common center" (Schelling 1994 (1797), 93)—a hub if you will—from which branched thought, history, and earth in a living principle of identity. Nature was not a world of mere appearances delimiting the reach of the ideal into the real, but a partner in the enterprise of human being and understanding. The partnership is ontological and originary—a matter of the "essential" (*wesentliche*) as it abides in the life of natural things. A remarkable advance at its time, this point of inquiry (together with his larger *Naturphilosophie* and theory of imagination) has also received great interest in the Schelling revival of the last decade, thanks to the efforts of scholars such as Jason Wirth, Sean McGrath, Fiona Steinkamp, and Joseph Lawrence.

In his early work Schelling inveighs against the mechanistic model of nature (then propounded in modern science and its enthusiastic focus on "matter") by insisting that, as Dale Snow summarizes, "the world of nature" is "an organic, living system" (1996, 69). There is a "purposiveness" coursing through the unity of form and matter, something revealing "a higher unity in nature than mechanism could ever reveal" (1996, 74). And this is precisely the terrain in which persons may come to understand themselves. Says Schelling, "The system of nature is at the same time the system of our spirit" (at Snow 1996, 81). In his 1798 *On the World Soul*, he warns against the trap into which the age of technique would come to fall: "as soon as I separate myself, and with me, all ideality from nature, nothing remains save a dead object and I cease to understand how life outside me could be possible" (at Snow 1996, 83). As soon as we conceive of nature as something analogous to a "machine," we veil its testament to a world whole and offend the way in which "nature is visible spirit and spirit invisible nature" (Snow 1996, 83–84). As he observes in 1801, "The power that bursts forth in the stuff of nature is the same in essence as that which displays itself in the world of mind" (Schelling 2001a (1801), 349). The statement, taken at its word, does not make nature a means to *theoria*'s ends. By "power" he means a grounds for the genesis of the real in nature's productive aspect and in the *will* that drives reason, something which philosophy had yet the means to conceive because it would involve recalling, say, the Cartesian *cogito* and the Fichtean self, back toward a more encompassing *physis*.

We can further draw Schelling into our figure of the threshold when we appreciate how he sought an absolute "ground" behind the divisions of subject and object, a ground that would present logically, yes, in the realm of intellectual intuition, and would present productively in the realm of nature. And what the pairing would mitigate, though he need not term it this way in his time, is the assumption that a glimpse of this ground must be had by entirely conceptual means. It would be had, rather, aesthetically, for the very life of grounding activity for spirit, for nature, and for the human imagination is something archetypally artistic. Although the controlling aspiration behind this idea might appear to encircle, rather than enter, the space between the "embers and the stars," it ultimately advances it. There is, he posits, a grounding "primordial whole, a 'harmony' between the ideal and real worlds that lives and moves as a productive and potentializing drive" (Schelling 1978 (1800), 12). In nature it consists in a nonconscious activity (the term is not disparaging); in the self it arises in conscious activity. But to "know" this primordial harmony—to grasp it in the act—requires an expansive form of what Kant had called *intuition*. A threshold must be crossed from the light of the conceptual thinking to the rich darkness of what could be called poetic sensing. Philosophical-poetic intuition, that is to say, may catch hold of its partner in *physis* because the two are built of the same stuff. But how so?

Schelling points to the model of the artist and the work of art—to what, in Kohák's Heideggerian vein, we have called the *poiesis* intended to be essential within *technē*. What the artist, like the philosopher, brings to his craft is what Schelling terms a special form of "genius" (here understood as "spirit" in a mode of highly concentrated creation). But unique to the artist—indeed *in* the artist—is a mode of nature's own ontological genius at a special pitch. Artistry is an outgrowth of aesthetic intuition, and the resulting work of art is, accordingly, a product of the living primordial harmony between self and nature. The artwork is "of" the threshold. Far more than a mere assemblage of form and matter, it is, says Schelling, a "miracle" insofar as it "radiates" back to us the root principle of identity between self and nature (Schelling 1978 (1800), 230). We misunderstand Schelling if we take all this simply as a way of assuring an absolute principle for the ambitions of Idealism vis-à-vis the "system" of reality. He intends, rather, to reveal how the aesthetic intuition of the artist can substantiate the potential for intellectual intuition on the side of the philosopher, for the artist has magnified how the productive power in nature can surface in the productions of the hand and the mind. To draw these points together, a certain "poetry" is unconscious but alive in nature, underway in aesthetic creation, and becoming-conscious in reflexive reasoning. There is, we might say, an inner dynamics going on at the threshold.

Schelling constructs a rich term for this production afoot in nature, art, and reason: *In-Eins-Bildung*. It signifies a creative "building-into-one," a unifying work of imagination that is underway behind the many particular scenes of what we may simply call "the real." That the attention to nature may pay off for the promise of intellectual intuition by no means makes light of nature. Schelling is, in his own pre-Kohákian way, calling the complacency of *theoria* to account. Nature, not mind, is the original source of art. Man's *Einbildungskraft* (power of imagination) serves at the pleasure of *Ineinsbildung*. "Connoisseurs and thinkers, however, because of their great inaccessibility of nature, for the most part find it easier to derive their theories more from the contemplation of the soul than from the science of nature" (Schelling 2021 (1807), 134). We might say, by way of contemporary parallel, that corporations and technologists, because of their heedless enframing of nature, for the most part find it easier to derive their warrant more from consumption than from cosmos. And like Kohák, Schelling is placing a question of "justification" before the bearings of human machination. Where Kohák furnished the basis for an answer by drawing attention back to the "sheer givenness of being" in nature, Schelling sets a course anchored in an aesthetic view of the world. There is a common *credo* between the two. As we have begun to see, Schelling regards nature and its entities in terms of a "living artistry" [*lebendigen Kunst*] which images the essential unity of thought and being (Schelling 1984 (1802), 125). The artistry, further, consists in a "striving" in nature and (drawing the lesson from above) in the artistic genius. Consider how far this is from the idea of nature as a mechanism and of thought as an apparatus of determinative cognition. The notion of *Ineinsbildung*, without diminishing their singularity, collapses the gap between object and subject and reclaims their kinship in a work that is the "innermost mystery of creation" (Schelling 2001b (1802), 386), something echoed in Kohák's "metaphoric sense" of reality. And as Kohák spoke of the dis-alienating possibilities inherent in realigning ourselves with the *Dichtkunst* of dusk, Schelling says that thought and Being enjoy their original identification in the living artistry of "the one-in-all and all-in-one" (Schelling 2001b (1802), 392). (Notably, the poet Samuel Taylor Coleridge transmitted Schelling's invocation of *Ineinsbildung* into English Romanticism in his 1817 *Biographia Literaria*. His neologism "esemplastic"—from the Greek *eis en plattein*, or "to shape into one"—for example, amplified Schelling's vision of the relationship between nature and the power of "primary" imagination.)

Schelling moves these matters further forward in an 1807 address called *Concerning the Relation of the Plastic Arts to Nature*. Imagine looking into an art studio set on nature's threshold. Building on his prior work, Schelling is troubled by what he perceives to be a persistent failure to think the theme of artistic production from the standpoint of a "living intermediary" position between soul and nature (Schelling 2021 (1807), 135). Artworks are not majestic on the basis of their "form," just as reason is not powerful on the basis of its concepts. He insists that "the basis of art and therefore beauty is in the vitality of nature" (154, n20). But most of us stand detached from "the holy and eternally creative primordial force in the world, which generates and actively produces all things out of itself" (135). Similar to Kohák's point about how human *cultus*, in its proper fit with earth, is best when teachable before nature's own "intrinsic sense," art is at its best when it "hosts" nature and the artworks draw forth their being from nature's wellsprings. The artist's own *technē* is to emulate the "spirit of nature, which acts within things," and artistic beauty "is being [*Wesen*], the universal, the look and expression of the spirit of nature dwelling within" (140). The stylized grandeur of these statements does not mean the points are naively romantic or simply aspirational. He has in mind a collaboration between nature's potency and man's creativity. Masters like Raphael, DaVinci, and Corregio understand how to do this, for when they "bring forth" [*hervorbringen*] their works they are enacting a "creative spirit" [*schaffende Geist*] wherein the "spirit of nature is freed from its ties and feels its affinity with the soul" (146). None of these achievements come about easily, but the highest works show the essence of art by virtue of their speaking "the eternal bond of a veritable divine love" together with an image of "eternal necessity" (148, 152). All of this amounts to a matter of striving in the poetic mode. And, as we saw earlier, *poiesis* is not limited to artistic production's relationship with nature; it is emblematic as well of "the primordial force of thought" (147). He singles out Raphael especially as a combination of these things, for in his works "He is no longer a painter. He is at the same time a philosopher and a poet," and "his work is impressed with the seal of singularity" (152–153).

What Schelling is arguing is that nature is the spring-like source of creative power, and that ascertaining art's essence requires "inspiration" in philosophy much the same as high art requires the striving inspiration of natural, creative force in the artist. The formulation of "gifts" that Kohák used to help convey the vital moral *sensus* issuing from nature serves well as a summary characterization of the work of *aesthesis* at Schelling's threshold. Stewards of *Ineinsbildung*, what artworks reveal is a creative bond between the eternal spirit and intelligence of nature and the properly inspired *technē* of the human imagination.

* * *

To speak with Schelling and Kohák of the "wholeness" of life and things, and the "grace" of the world is by no means to make light of the assaults we have made on the earth. Nor does it make much "practical" or policy headway in repairing the damage. But the visions they pursue offer a formative intervention in the way we stand at nature's threshold. They turn us from a decadent art of the dominant to an original and constitutive artistry of hospitality. With his generic naturalism, Kohák reintroduces us to the abiding domain of nature's *moral sense*. He envisions a praxis that would apply the question of "justification" to our *technē* and ourselves. With his naturalized Idealism, Schelling reintroduces us to the originary domain of nature's *aesthesis*. He celebrates a praxis of retraining our creative impulse—in art and in thought—on terms set by the life of a shared primordial whole, a common "center." Both in their own ways ask, *Who are you who live in all these many forms?* with an openness to the *physis* in nature and a critique of the ways of *anthropos*. Venturing the threshold experience, their paths cross one another in an important way. Kohák's kenotic step into the task-granting grace of natural revelation traverses Schelling's creative step toward the poetic charge of a living system. The horizon forming around their convergence amounts to a topological encounter with what is native to earth but too long masked or misappropriated by us: an art of the possible.

Notes

1 See especially Richard Kearney and Melissa Fitzpatrick's wonderful exposition of these phenomena and the ethics of the 'host/guest' relation at the 'threshold' in *Radical Hospitality: From Thought to Action* (Fordham University Press, 2021).
2 For the Heideggerian background to Kohák's critique of technology, see Martin Heidegger, *The Question Concerning the Essence of Technology, and Other Essays*, translated by William Lovitt. New York: Harper & Row (1977).

References

Heidegger, Martin. 1977 (1954). *The Question Concerning Technology, and Other Essays*. Translated by William Lovitt. New York: Harper & Row.
Kearney, Richard. 2010. *Anatheism: Returning to God After God*. New York: Columbia University Press.
Kearney, Richard, and Melissa Fitzpatrick. 2021. *Radical Hospitality: From Thought to Action*. New York: Fordham University Press.
Kohák, Erazim. 1984. *The Embers and the Stars: A Philosophical Inquiry into the Moral Sense of Nature*. Chicago and London: The University of Chicago Press.
Kohák, Erazim. 2000. *The Green Halo: A Bird's-Eye View of Ecological Ethics*. Chicago and La Salle, IL: Open Court.
Malick, Terrence, director. 1998. *The Thin Red Line*. Fox 2000 Pictures & Phoenix Pictures. 2h, 51min. Released in the United States on January 15, 1999. www.imdb.com/title/tt0120863/
Schelling, Friedrich. 1978 (1800). *System of Transcendental Idealism*. Translated by Peter Heath. Charlottesville, VA: University of Virginia Press.
Schelling, Friedrich. 1984 (1802) *Bruno, or, On the Natural and the Divine Principle of Things*. Translated and edited by Michael G. Vater. Albany, NY: SUNY Press.
Schelling, Friedrich 1994 (1797). "Treatise Explicatory of the Idealism in the Science of Knowledge." In *Idealism and the Endgame of Theory: Three Essays by F.W.J. Schelling*. Translated and edited by Thomas Pfau. Albany, NY: SUNY Press.

Schelling, Friedrich. 2001a (1801). "Presentation of My System of Philosophy." Translated by Michael Vater. *The Philosophical Forum*, vol. XXXII, no. 4 (Winter).

Schelling, Friedrich. 2001b (1802) "Further Presentations from the System of Philosophy." Translated by Michael Vater. *The Philosophical Forum*, vol. XXXII, no. 4 (Winter).

Schelling, Friedrich. 2021. (1807) "On the Relationship of the Plastic Arts to Nature." Translated by Jason Wirth. *Kabiri: The Official Journal of the North American Schelling Society* vol. 3: 132–158.

Snow, Dale E. 1996. *Schelling and the End of Idealism*. Albany, NY: SUNY Press.

Chapter 11

Rethinking a Hospitality of Nature: Three Colloquies

David Storey

Let me say a word about the title of the panel series excerpted in this volume, specifically the words in the title. Take "ecology" first: the word "ecology" has become commonplace in our cultural lexicon, but this was not always so. It was coined by the German biologist Ernst Haeckel, often referred to as the "German Darwin," in the late 19th century. He defined it as, "The study of the relation of the animal, both to its organic as well as its inorganic environment." Ecology has, of course, grown since then into its own science, or, more precisely, its own family of sciences, and become a prevalent way of thinking about the natural world and our place in it. But, then there's a second word: "wisdom." A hundred years ago, poet T.S. Eliot wrote, "Where's the wisdom we have lost in knowledge? Where is the knowledge we have lost in information?" Ecological knowledge acquired through the natural sciences is essential for thinking through our ecological crises. But science alone is not sufficient. It cannot tell us what we ought to do with that knowledge, how we ought to live, what sort of society we want to build, what sort of world we decide to pass on to our descendants. We also have our wisdom traditions: philosophical and religious, Eastern and Western, indigenous, ancient, medieval, modern, and postmodern. And our challenge is to retrieve and integrate the best of them in our own time. We often understand ourselves as living in a knowledge-based economy. As it happens, our word "economy" shares with ecology the same root, meaning "household." Our task now is to rebuild our economies in a way that aligns with and respects our ecologies so that they better reflect the original meaning of both terms: the laws and logic of the "home"; call it a "wisdom economy."

Panel #1: Earth Justice

We traditionally understand justice as an exclusively human affair. But the ecological crises—there is not one, but many—demand that we expand our sense of justice to comprise not just humans distant in space and time, but other species and, perhaps, the earth itself. Doing so leads, our panelists argue, to a rethinking of *who* we are, *where* we are, and *when* we are. We must come to see ourselves, and reshape our world, to reflect the fact that we are both generational and geological beings, ancestors and descendants, guests and hosts of the earth itself.

DOI: 10.4324/9781003456940-15

Olúfẹ́mi Táíwò argues that in order to pursue climate justice, we must adopt "the ancestor perspective." To do so, we must first understand ourselves as descendants, and come to terms with the world we have inherited, a world shaped by what he calls "global racial empire," built on the colonial project and the trans-Atlantic slave trade. While slavery and colonialism are formally over, their legacy persists not only in terms of the resource disparities between the global North and the global South, but in both the skewed vulnerabilities to climate change and the capacities to adapt to it. He argues we should see this world as, fundamentally, a distribution system not merely of resources such as wealth and food, but of opportunities and affordances such as self-determination and democratic equality. Drawing on the "capabilities" approach developed by scholars such as Martha Nussbaum and Amartya Sen, he argues that climate justice must be an exercise in "world building" to more equitably distribute human capabilities. Such an intergenerational construction project must build tomorrow's world in light of yesterday's injustice.

Matthias Frisch similarly thinks that climate justice requires us to remember that we are intergenerational beings. Frisch agrees with the emerging discourse around the Anthropocene that we must situate human history within geological deep time but argues we must go further. We must understand ourselves not merely as *geological* agents, but *generational* agents. Frisch's major idea of "taking turns with the earth" has a double meaning: each generation *takes a turn* with the earth—using and stewarding its resources for the next generation—each generation takes a turn *with the earth*. Put another way, we not only take turns with the earth, but the earth takes turns with us. This latter sense involves a radical rethinking of the earth system. With the collapse of the Great Chain of Being, the Copernican Revolution, and the rise of Cartesian dualism, we became accustomed to seeing the earth as a physical, celestial object and a stock of resources. The fantasy of building bases on Mars after we have exhausted Earth's resources is a sign of how this view of the earth is a legacy of the frontier mentality fueling the colonial project. In contrast, Frisch draws on what is sometimes called the "second Copernican revolution" associated with scientist James Lovelock's Gaia Hypothesis. We are not so much "on" the Earth as a lamp sits on a table, as "of" and "in" it. It is constitutive of our humanity—as he points out, the very word human is related to hummus, dirt. The Earth, properly speaking, is not a planetary object, but the "critical zone" spanning the sun, atmosphere, biosphere, hydrosphere, and lithosphere that supports life.

Stanley Anozie draws on a wide range of thinkers—including Martin Buber, Martin Heidegger, and Alan Gewirth—to challenge the anthropocentrism that afflicts our conventional thinking about justice. He expands Gewirth's concept of a "community of human rights" beyond a humanistic scope to encompass the nonhuman world. This expansion is grounded, in part, upon the idea that nonhumans have agency and purpose in their own right. We can and should relate to them not merely as objects in causal interaction, but as subjects in communication. To use Martin Buber's language, our relationship to them is not merely that of "I-It" but also of "I-You." Like Frisch, he traces the denial of the interiority of the nonhuman to the legacy of René Descartes and proposes instead that all being is "being with." Nonhuman beings are not simply complicated machines moving around on the earth or in ecosystems, but they are members of a biotic community that is their home. Taking a page from Pope Francis, whose integral ecology asks us to recognize "Brother Sun and Sister Moon," Anozie invites us to see ourselves not as the pilots of spaceship Earth, but as members of a common home.

Panel #2: Nature in Asian Traditions

Our second panel is entitled, "Nature in Asian Traditions." Graham Parkes is one of the world's leading scholars in comparative philosophy and a celebrated translator of seminal works by Friedrich Nietzsche and Nishitani Keiji, but more recently he has turned his attention to the climate crisis. Here, he sets out to explain why, in his view, human beings have become such poor guests of the Earth. His answer is that we have forgotten we are guests in the first place, that the earth is a home. Underlying the ecological crises is a view of the natural world that primes us to treat it without respect. Many defenses of the environment focus on the value of and respect for the animate world of animals and plants, but Parkes focuses on the overlooked strata of so-called inanimate things—both natural and human made. In his telling, the devaluing of matter began with the Christian positioning of God outside of nature and reached its apex in the 17th century, when natural philosophers such as Descartes and Bacon came to see the material world as dead, inert, spatially extended stuff. Alfred North Whitehead's description of scientific materialism captures this view: "nature is a dull affair, soundless, scentless, colorless, the mere hurrying of material, endlessly, meaninglessly." This view, Parkes points out, is an outlier. For most of human history, people regarded the natural world as something alive, animate, and ensouled, and this view found more articulate expression in the philosophies of Ancient Greece as well as ancient Asia. Parkes points to Daoism and Mahayana Buddhism as providing the richest expressions of how matter matters; in Daoism, all things are constituted by and expressive of *qi*, life energy, and in Mahayana thinkers such as Zhanran, Kukai, and Dogen, all things are regarded as brimming with Buddhanature, including artifacts. By befriending things in our immediate environment, we can develop the mindful reverence needed to deal with our global predicament.

While Parkes's argument focuses on changing how we view things by looking to inspiration from Daoism and Buddhism, leading environmental ethicist Marion Hourdequin's centers on changing how we view persons by drawing on the third major East Asian tradition: Confucianism. As she points out, many of the environmental challenges we face, from climate change to plastic pollution to nuclear waste disposal, are intergenerational. Are the moral traditions of the societies that caused these problems sufficient to solve them? In particular, is the individualistic ethos that characterizes Western thought fit for purpose? Hourdequin thinks not and suggests that Confucianism offers a useful alternative. In this tradition, persons are thought of as "persons-in-relation." Relationships are not caused or voluntarily contracted by persons; rather, relations constitute personhood. The Confucian credo might be termed, "I inherit, therefore I am." Both to exist, and to develop into a mature human being, I depend on a web, net, or chain of relations extending back into the past and forward into the future. This view of human nature generates distinct forms of gratitude, reciprocity, and rationality. While the Confucian tradition is generally humanistic in scope, Hourdequin argues that its ethos has obvious relevance for today's environmental predicaments.

Like Parkes, Leah Kalmanson engages the challenge of re-enchanting the world, but in a specific context: indigenous Hawaiian religion and a legal controversy over the ownership of a local plant species. The philosophy of religion has long had trouble finding language for such traditions, having inherited a schema assembled in 19th century scholarship that distinguished between the "great" religions—the monotheisms—and the "not so great"—those found in Africa, the pre-Colombian Americas, and Oceania, which are deemed "primitive," "tribal," "pre-literate," and so on. Like Parkes, she thinks we should resist the modern reflex that brands such worldviews as naively animistic. She draws on the work of Freya Matthews, whose book *For Love of Matter* was an early entry in today's revival of panpsychism. She sees in Matthews' metaphysics a resonance with Chinese cosmogonies that posit a movement from a primordial one—emptiness or undifferentiated chaos—to two—yang and yin, the polarity of opposites—to the "10,000" things, i.e., the many of the manifest world. Such a cosmology expresses the underlying unity of all things, yet preserves their particularity and, important for Kalmanson's purposes, the uniqueness and autonomy of cultures and traditions. To properly relate to indigenous traditions such as the Hawaiian case, we must be careful not to fall prey to the colonial reflex to assimilate the Other into the Same. In trying to be good hosts to the Earth, in other words, we must learn to be good hosts to her other, older, and in some cases better hosts.

Panel #3: Ecological Endgames

Ecological discourse is rife with the language of apocalypse. For many, the human game is at or approaching an end, playing itself out as we destroy our common home. But a different way of thinking about—and playing—endgames is to think about the end of the human game, in the sense of its *telos*—its point and purpose. Whether human civilization proves to be, to use the theologian James Carse's distinction, a finite or an infinite game, depends on how we understand and enact the human–nature relationship in the coming century.

This conversation is something of a scholarly reunion. Ariel Salleh and Michael Zimmerman first met nearly a half century ago at the first international conference on ecofeminism, and both were present at the creation of environmental philosophy as a field of scholarly inquiry and research. Here, they draw on their respective backgrounds in ecofeminism (Salleh) and Heidegger and deep ecology (Zimmerman) to discuss the history and status of environmental thought and practice, and the ethical complexities of a range of issues including nuclear power, renewable energy, population dynamics, and the compatibility of capital-ism and a sustainable planet.

Zimmerman's entry into the field came through his study of Heidegger's critique of modernity. For Heidegger, the essence of modernity consisted in the advent of the technological way of thinking (*Gestell*). Technology is not mainly or merely tools and machines; it is a way of making sense of the world that reduces things to a storehouse of energy (*Bestand*) for human use and consumption. According to Zimmerman, Heidegger traced this "productionist metaphysics" all the way back to the Greek philosophers, whose concepts were closely modeled on handi-craft production. To be is to be produced. In the Medieval era, this morphed into the notion of nature as created by God, but in modernity, humanity became the measure of all things, not only the "masters and possessors" of nature, as Descartes put it, but also increasingly the *producers* of nature. As Bill McKibben put it in the first popular book on global warming, we have reached the "end of nature" as a wild world untouched by human activity.

While Heidegger sought a third way beyond communism and capitalism, both of which he saw as caught in the matrix of modernity, Zimmerman points out that his philosophy of nature was shaped by and entangled in the politics of National Socialism. We typically associate environmentalism with the post-war Left and, indeed, conservatives lampooned the greens as "watermelons"—green on the outside, red (communist) on the inside. But Zimmerman explains that between the two world wars, a powerful environmental movement flourished in Germany, albeit an ethnocentric one suffused with the ideology of "blood and soil." As climate change intensifies, producing environmental refugees, disrupt-ing agricultural yields, and fanning the flames of xenophobia, racism, nativism, and nationalism, we must be wary of what Nils Gilman has termed "avocado politics": green on the inside (regimes that accept the realities of climate change), but brown on the outside (using the climate crisis to secure access to dwindling resources through armed conflict and foment hatred or violence toward immi-grants and cultural minorities). Heidegger offers us both a critique and a caution-ary tale of dark ecological endgames.

Ariel Salleh came to environmental thought through her work on eco feminism. As philosophers began to rethink the relationships between human and nature in the 1970s, Deep Ecology emerged as an intellectual movement aiming to resist the thoroughgoing anthropocentrism of Western thought, drawing on thinkers and traditions as diverse as Buddhism, Hinduism, and Spinoza, as well as eco-logical thought itself. Many, including Zimmerman, drew connections between some of modernity's strongest critics, such as Heidegger, and the Deep Ecology platform. Salleh was among the first thinkers to challenge this way of thinking under the banner of ecofeminism. To the deep ecologists' critique of the subordi-nation of nature to humanity, the ecofeminists added the subordination of women to men. As Salleh puts it, ecofeminists are intent on challenging the binary "one or zero" logic that they claim runs through Western thought, or what Karen Warren called "the logic of domination." This logic splits reality in two—humanity/

nature, human/animal, man/woman—and erects a hierarchy subordinating the latter term. This logic, they claim, is at work in capitalism, colonialism, and in our relationship to the Earth as such. To replace this destructive and dehumanizing logic, we must find new models, such as indigenous ways of knowing and living, to build what environmental activists Vandana Shiva and Paul Hawken have called a "regenerative" civilization.

EARTH JUSTICE
Speakers: Stanley Anozie, Olúfẹ́mi Táíwò, Matthias Fritsch, David Storey

Olúfẹ́mi Táíwò: I want to begin by saying something about the issue of "reparations" as it relates to Climate Justice. More specifically I want to raise the idea of "reconsidering reparations," which concerns the perspective of ancestors. Why is it helpful to think about ourselves as ancestors when we're thinking about climate justice and the various challenges of it?

Something I've been discussing in my recent work are these two contentions: one, that reparations require remaking the world and, two, that the nature of that challenge in our particular time period should lead us to thinking about climate justice. Reparations, racial justice, and climate justice are things that ought to converge in terms of concrete goals and aspirations. To say a little bit about why I have that perspective: I understand reparations as world-making, because I understand the era that reparations respond to—at least reparations for trans-Atlantic slavery and colonialism—as being the era of the construction process of the world. And the aspiration for racial justice on a planetary scale is not new; it's not something I'm introducing by any means. In fact, Adom Getachew's recent work argues that that scale of aspiration was in fact the prevailing way of thinking about struggle in a previous time period in the wave of anti-colonial movements that followed the Second World War. So, it's not new by any stretch, but it does come from a particular way of thinking about the world.

Broadly speaking, I'm committed to this way of thinking concerning world history over the last few centuries. We have a global social structure now, and that global social structure was built, was constructed, by trans-Atlantic slavery and colonialism. And because the world was constructed when it was and in the way that it was, the world that we have now largely has some unjust features. Social advantages tend to get produced and distributed across the world in a way such that they accumulate for global Northerners, for the racially advantaged. Disadvantages tend to accumulate in the global South and for the racially disadvantaged. And this seems to hold at multiple scales of analysis: whether you're thinking about within countries, or whether you're thinking about across countries; we're comparing countries.

So, the thing that I argue for is that the right way to respond to a history that produced that kind of world is a construction project. What we should do is justly distribute the cost of creating tomorrow's world in light of who's accumulated what and after yesterday's injustice. There's a lot of things that we could think about if we're thinking about the world as a distribution system. We could think about where resources are, maybe, well, dollars, or the kinds of things that we're used to thinking about. And those are certainly very important resources that we should think a lot about.

But the view that I like is called the "capabilities perspective." What is important, what we need to distribute are the parts of the world that allow people to do different things. We could distribute, we could shape the world in such a way that people have the ability to be fed by distributing dollars so people can buy food. But we can also distribute fruit trees, right, so people are near sources of food. We could distribute jobs. There are lots of approaches. What's important is the outcome, that people are able to eat, and so on and so forth.

I'm centrally concerned in this presentation with team capabilities. The point of doing that is to let people have self-determination, to be able to determine the course of their own lives. The way that we should think about distributing things in order to give people self-determination is to distribute affordances so we can actually build the world in a way that makes it usable by different people in different sorts of ways. And here I'm taking lessons from disability justice activists, like those who were part of the "Universal Design" movement. It's not well-described in terms of resources. If we're thinking about the problem of who can access which building, it's literally what the building is built like: are there stairs in front of it or are there ramps in front of it? Those are the things that are going to determine access, and that's the way that we should think about access to buildings. That's the way that we should think about justice, at least in its design, whether we're thinking about the design of literal physical buildings, or whether we're thinking about the design of institutions, and so on and so forth. We should design our political institutions and our buildings in ways that learn from this movement. These were the principles of design rather than simply moral principles.

But all of this sets up a pretty big problem. We have to change a planet-sized political system, and if I'm right, that means we have to *literally* rebuild a planet-sized political system: change physical structures, change political structures. If we could do that at all, the timescale for changes, the magnitude of that kind, seems like it's probably quite large, quite long, right? It might take many generations to get something like that done. But we only have so much time for climate justice. We only have so much time before we need to respond to political problems in our own lifetime. How do we square the circle? How do we respond to the huge timescale for changes of the magnitude that we're looking for? The thing that I say is that it may be mysterious concerning *how* we could do this, but it is not mysterious *that* we could do this, because we already do.

The example that I use in my book on reparations is about making soy sauce the old way. Yasuo Yamamoto is a person who does this. Generations ago, his grandfather and a neighbor planted cedar trees which grow over many decades and provide the wood that makes it possible to make soy sauce in this old way. Not just that, but somebody else who is a contemporary of Yasuo Yamamoto, a craftsman, Takeshi Ueshiba, worked with a team of people to make the barrels that Yasuo Yamamoto used to brew the soy sauce. All of that is to say that there were some contributions from long ago made by his ancestors by planting trees, some contributions from other people that are nearby in timescale, such as those made by Ueshiba and his team, and some contributions from the weather and from microorganisms which happen over huge geological timescales by making any particular weather patterns or by making any particular climates reliable. Yasuo Yamamoto in fact succeeded and was able to brew soy sauce in the old manner because all of these things got together.

It seems like we do rely upon each other on these huge timescales and, in fact, act like ancestors with respect to each other. But Yamamoto got lucky, because the world isn't necessarily built around making that kind of huge timescale collaboration possible. It's built around other things like, say, maximizing shareholder value. So, all that I propose to answer this with is just the constructive view of reparations. It's just leaning into the notion of reparations. What we need to build are stable patterns of interaction—institutions, food systems, housing systems—that are reliable on those large timescales. It's a thing that comes right out of the capabilities approach, and all this means is that we should think about this differently. We shouldn't think about this all in one fell swoop like we must right now build the just world, as if the only moment for change is *now*, because this process is much longer than can be addressed only in the present. But we should be trying to think about building the sorts of things that are going to get us closer to the just world, that are going to put our descendants in a better position to get there. We have some moral guide-posts that are also design principles: democratic equality, self-determination, or letting people decide their own lives. And we can, right now, pick ways of relating to our political possi-bilities that expand those and that expand the possibilities for the people who come after us to do even better.

Matthias Fritsch: The theme of this event explores the question of ecological hospitality by asking what it would mean for us to be guests of the earth as well as hosts of it. I have some questions about what "us" here means, but maybe we can get to that later as well. And I particularly like the idea of being guests and hosts with the earth, and this, I think, gestures toward a kind of reci-procity view when it comes to nature or the earth that looks to include the ideas of receiving from and giving to the earth. And that is very important in my account of intergenerational justice as well, and in my account of climate justice. But here I discuss reciprocity with the earth as human generations "taking turns" with the earth and with each other by pointing out that the earth here has some agency and also turns generations about. I want to indicate a little bit more about what that means.

I want to link my account of intergenerational justice as "taking turns," in particular taking turns with the earth, with the recent proposals for what's sometimes called a "geoki-netic" view of the earth, or the second Copernican Revolution. And I'll talk about that toward the end. First, I want to make the point that recent proposals that suggest that we should be situating ourselves in geological time—so, very, very, very long timescales, millions and millions of years—state that issues dealing with climate change and the Anthropocene requires us to situate ourselves in this way.

I want to say that that's actually not very helpful when it comes toward accounts of climate justice. What we need are accounts of justice that are global in scale because we want to pay attention to the most vulnerable today, but we also need generational time. We want to pay attention to the vulnerabilities that come with future generations that are much more proximate than these long geological timescales. But, that requires that we conceptualize intergenerational justice, and here is where I briefly review my much longer account of taking turns with the earth.

I want to say that the earth here does have some agency, as I indicated earlier. It's not just an external exchangeable object, but it is constitutive of us and, because it is constitutive of human beings, we can say that earth also takes turns with us. And, to flesh this out, I turn to this idea of the second Copernican Revolution. Here are just some ways in which people have argued that we should be situating ourselves today in deep ecological time. David Wood, Dipesh Chakrabarty, various people, have been arguing this, so we need what's called "deep time" or we need to be "geologically human." So, we should understand ourselves as emerging from very, very long timescales that have made the kind of conditions possible in which human lives and human capabilities perhaps can flourish today. So, we should understand there to be this kind of dependence of these long timescales.

There are various worries about this notion of the Anthropocene and of deep time. One of the big ones, of course, is that we should think of humanity as geographically and socioeconomically divided. It isn't just a question of humans in general owing themselves to these long timescales, but that we need to take into account that some humans profit a lot more from these conditions than others and, in particular, as Olúfẹ́mi mentioned, the difference between the global North and the global South in this context is very important. The differences between classes are also important.

But what I want to focus on here is that this notion of the Anthropocene conceptualizes the human species not just as spatially divided (which is a point that's often made in these discourses), but also that we should think of the human species as temporally differentiated, namely in terms of generations. And part of the reason why I'm worried about this notion of the Anthropocene is that it tends to focus the concern—I see this again and again—on human survival. So, really, regarding the climate change, environmental destabilization, and so on, that is really when it comes to the future, that's a problem, because humanity is at stake, human survival is at stake.

I find that very worrisome. One, because only on quite unlikely scenarios is that actually a real worry—the Intergovernmental Panel on Climate Change think that's very, very unlikely. There are some projections that indicate that mammalian life in general may be in danger, but those are unlikely, and we shouldn't be focused on them. I think that we should be focused on, when it comes to forward-looking concerns, we should be focused on this spatial-geographical injustice that I mentioned earlier, but also we should be focused on intergenerational injustices. Part of the problem with making the survival of humanity the key goal for looking toward the future is that it privileges the rich once more, I fear, namely that they can say, "Well, I need to build this bunker because I'm human and I need to survive; I need to carry on the torch of humanity," or something like that, right? And that's what we don't want. I think you'll find this in debates between Andreas Malm and Dipesh Chakrabarty, for example. So then the pro-futural concerns should not be understood primarily in terms of human survival, but they should instead be understood in the sense of intergenerational justice.

And then that requires that we conceptualize generations and generational justice by talking about generations taking turns, something that I have tried to do. And regarding the notion of generations taking turns, I basically argue that this is something we are doing. Generations, in fact, are taking turns with the earth. We always inherited the earth from ancestors, and we are passing it on to future generations. This is something that we do; it's a fact, but it's also a fact that's rich with normative possibilities, because we can do this in better or worse ways. I think the taking turns idea is very interesting in that it's situated at both this social-ontological as well as this normative level, where you can then try to flesh out the normativity that's contained in taking turns, right?

Think of kids on the playground: they take turns with a swing, and that is a fact, but you can also then ask, "Well who of the kids there is taking a better turn? Who is really stepping down when it's not their turn anymore, or who is not damaging the swing when they're doing this?" and so on. So, that is part of why I like the idea of taking turns. So, that's basically what I'm trying to say here, and a lot can go into asking the question of what it is to take a fair turn, which, I think, should be the question of our time when it comes to climate and the earth. What is it to take a fair turn? One of the things that I like about it is that it is so much part of our socialization that, wherever you go, people have ideas about this, right? Because so much of a part of language learning is that you have to wait your turn when someone else is trying to speak. Board games rely on this and, again, with kids waiting for the swing, and so on.

It's better than the other model—the much more dominant model, especially in western moral and political philosophy—where sharing something is sharing by parts, as if one were cutting it up, like a cake-cutting model. This is seen in Rawls, for example, and in many others. So, distributive justice is typically thought of in terms of sharing by parts, and I think sharing by turns is, when it comes to things like climate and the earth, much more useful.

But the problem with the taking turns model, I think, can be... one of the problems of this view can be expressed in the sense that it makes the earth or the climate as something that seems external to us. So, when we think about the swing example, or when we think about the bicycle example, I don't have to use the bike. The bike is a choice: I can walk or I can take the bus, or something else. So, the earth is not like that. The earth is different; we don't have another one. There is no other choice but to live... for generations, to live on and of the earth. And so, this is something that I really want to flesh out more, that is, regarding what may be the difference there. And one way to talk about this is to say that the earth is in fact constitutive of the turn-takers. The earth is that which also gives birth to the turn-takers. Human generations are not just born of a previous generation, but they are also born of the earth. And that is what makes a difference there, compared to the swing or the bicycle (or what have you).

The earth also takes turns with generations and does so in various ways. For example, the earth takes turns with generations by disrupting the unity of a generation. It disrupts the unity of a generation because what earth does in a certain way is that it gives life, but it also takes it. But, it takes it individually such that one always has the problem of defining the point where a generation begins and where it ends because there's this constant flow of births and deaths. And regarding this, I think of it as the earth sort of disrupting the linear flows of what otherwise could be perceived as linear flows of (many distinct, unified) generations.

So, the earth takes turns with generations, but that is something I'd like to flesh out a bit more with the recent proposals for a Counter-Copernican Revolution (aka the second Copernican Revolution), or what I earlier called the "geokinetic" view of the earth. You find this position in various authors' work, such as Michel Serres, Bruno Latour, Thomas Nail, and others. So, what is the second Copernican Revolution? It shows that, on long timescales (there we have them again), earth also displays an internal movement. The first Copernican Revolution stressed that the earth is on the move, but, externally, among the turning around the sun, and so on. But the second Copernican Revolution emphasizes that there's also an internal movement that we need to take into account. That is now what we see with all these environmental destabilizations, with the various things that are going on; climate change is only one of them. And this shows us that while the earth in fact has always been on the move in a certain way, it has always had some agency; it has been internally constituted in such a way that it can be constitutive of human generations and can also disrupt them and can create conditions for flourishing and can lower the conditions for flourishing as well. And that's what we have to take into account.

The first Copernican Revolution usually went along with what Latour called a deanimated view of the earth or of nature, and an overanimated view of culture or of humanity. The way that agency was distributed was that humans got all of it—or some humans, in fact, got a lot of it—and the earth got none of it. It's all just that matter out there that's available for resources, extraction, and so on. And the second Copernican Revolution must undo this very modern distribution of agency to grasp the massive mobilizations on the earth that we see today. So about half of the species are on the move today as a result of climate change and various other forces and various other destabilizations of habitat. There's a massive movement of species that we see now. And there's climate change, ocean acidification, species extinctions, and so on. Mass planet and animal movements (and so on) show us that the earth system is internally on the move, and in fact was always capable of this. Only now these environmental destabilizations grasp our consciousness, only now they are of great concern. They are happening in a much quicker and more threatening way than before.

Now, this means we have to think of the earth differently than in this way. Look at all the images displayed, e.g., by the UN, which show the earth behind held by human hands. These views, I'm trying to say, are wrong, they're mistaken. They suggest that the earth is like the bicycle, like the swing. And that's not the way we should think about the earth, because what we're missing is this constitutive nature of the earth, that the earth is, in fact, that which also gives birth to us, and also withdraws that birth in a certain way. We're finite, we are mortal, and when we die, we pass back into the earth and its internal movement. So, we will be decomposed whether we like it or not; the earth will recompose our remains, right?

Latour and others propose that we think of the earth… that we flip the earth. It's only here on the outside, on this very thin layer that's called a "critical zone" by scientists. The critical zone is where life is actually possible, right? But, we should think about it in such a way that the critical zone is actually on the inside and not on the outside. I think that both this view of the earth where you can map it out as a ball, where everything can be seen from the outside, as well as the view of the earth that we had in the images of it being passed on, as it were, where the hands are holding it, are obviously problematic.

But, that is also connected to this idea of extending the frontier. You have in the history of colonialism and imperialism all these mappings-out of the earth, especially by European powers. And the idea with that is where can we go next, what can we colonize, what can we occupy? So, this mapping of the space of the earth I think very much belongs to a colonial mindset where we need to extend the frontier and go elsewhere. And that is also what we now find with colonizing (outer) space, right? You see this in all these various ideas that say we need to find exoplanets where life is possible and that we can colonize, right, that say we need to colonize Mars, we need to set up colonies there. It's still the same modernist mindset that doesn't quite accept that the earth is constitutive of us, I want to say so. But the earth is constitutive of us, but in such a way that we cannot but leave it behind for the next generation, right? The earth constrains us to leave it behind or to take turns with it because we cannot keep it in our hands. And that is what I think this geokinetic view of the earth, or the second Copernican Revolution, shows us so well.

So we owe the earth to the next generation, to the next turn-taker, because we are of the earth and are not contingently placed on the earth, right? We're living on its inside and not on its outside, if you want to use those metaphors. And so, that I think is the way in which intergenerational justice and the reciprocity view with the earth are connected. Humility is also a word that comes from *humus*, which means soil, or earth, or dirt, right? So, stay close to that, and so that's why I sum it up by saying *tour-à-terre, terre-à-tour*, right? We take turns with that with which we are grounded in, literally grounded in.

Stanley Anozie: I would like to focus my attention on the community of rights and earth justice. You may call it environmental justice or ecojustice. My line of argument will be from the tradition of Alan Gewirth, especially as found in his work *The Community of Rights*, and then also from the idea of "right" by delineating an understanding of rights or human rights. My intention is to use the principle of generic consistency. I want to go beyond an Anthropocene interpretation of reality, which is about human supremacy and humans being the center of it all and about human survival, by pointing out that we are, just like Martin Heidegger would talk about, *geworfen*. The idea of *Geworfenheit* is that we are thrown into the world, we are part of this universe, we are constitutive of the world, and the world is constitutive of us. And so, regarding the idea of Alan Gewirth's principle of generic consistency, I want to look at it in a way where we are no longer being generic on ourselves as agencies or moral agencies or human agencies (to the exclusion of all that is non-human), but we are instead looking at nature, the environment in its totality, human included, where every other fate is included as part of it.

And so, the world, nature, the universe, is not going to be conceived from René Descartes' perspective that says that we are the masters and possessors of the universe. The universe is not a space to be conquered, or, like Matthias pointed out, it is not a colonizable space. And we are not living authentic existence by looking for spaces to conquer and populate in. It is about the space of authentic being, and the space of recognizing the otherness of the other. And when I say the otherness of the other, I'm not just talking about another human being. I'm talking about the otherness of nature, the "it" part of it and the "they" of the human part of it, the animated and non-animated aspect of it.

Now, I would like to align this principle of generic consistency by Alan Gewirth to an idea of Martin Buber's, to the idea of the interpersonal relationship; not just the interpersonal in terms of human being to human being, and also not just the relationship of I and Thou, but the authentic relationship from the perspective of I-it. And this gives me this idea of these two German concepts that come to mind: the ideas of *Vorhandenheit* and *Zuhandenheit*. *Vorhandenheit* is the nature, the reality of things, animate and inanimate; they are present to us. This presence does not imply they are being conquered, colonized, dehumanized, or used solely for our own purpose, for our own well-being. It is about a constitutive, collective, generic well-being. That is, it is about our well-being, our authentic existence as well as the continuity of the world, or the continuity of nature itself. And then the other aspect of it is the idea of *Zuhandenheit*, which is the German connotation of this usage of nature, but this usage doesn't necessarily have to be about abuse. It has to be authentically about using the world or the things of life as they are to be handled because they have their own agency. I use "agency" in a generic sense: "agency" as in "purposefulness," but that can also be related to the idea of Aristotle's final cause, the goal of a thing. And if that is the case, it means that the agency I'm talking about here is not an exclusive capability or capacity of the human being. It is a capacity, a feature, an authentic and constitutive aspect of every aspect of nature, which means trees have their agency and purposefulness, birds have their agency and purposefulness, the sun has its own agency and purposefulness: every aspect of the world has agency and purposefulness.

And then, the other point that I find very interesting in this deliberation is to emphasize what Immanuel Kant says that I think is connected with the idea of this notion of *Geworfenheit* by Martin Heidegger: we are thrown into the world, and the world is our space of being-with other people, and not just other people, but also being-with animate beings and non-animate beings, as the case may be. And we are not strangers in this space, either, we are not strangers of the earth, we are not "owners" of it, we are not "masters" of it. This even helps me to make some references with (I think it was Pope Francis talking about this) the Brother Sun and the Sister Moon.

So, you see that the animation of the inanimate aspect of our existence is this understanding that we are, together with other creatures, together with all the aspects of the word of "nature," thrown into this space, this space of existence. We have to recognize every being in every aspect of nature as having a home. We are not hosts, we are not guests; we are not lords, we are not masters. We are an essential part of it. The birds recognize us, they recognize our agency. The trees recognize our own agencies, but they may not have the human language that we have that we use to express our own understanding of the world from, so to say, an anthropocentric perspective. In other words, my position is to emphasize that man, or human being, is not this measure of reality, nor the center of reality. Nature is reality, reality is nature. I think this resonates with the position of someone like Baruch Spinoza, where he says "*Deus, sive Natura*," "God or Nature." And so, if that is the case, the inclusivity of nature, the constitutive nature of nature, implies that we have human beings, but also we have all the beings, including non-human beings or non-rational beings. These latter beings have this generic agency that we ought to recognize and to live with if we have to live in a world of authentic existence, which is a world of earth justice and a community of right.

NATURE IN ASIAN TRADITIONS
Speakers: Leah Kalmanson, Graham Parkes, David Storey, Marion Hourdequin

Graham Parkes: Human beings are winning the Earth's Worst Guest Species Award, hands down. Not only are we annihilating increasing numbers of other species, but we're also on course to destroy enough of the Earth's biodiversity to bring about our own annihilation. But, our treatment of human-made things is also a problem here, as evidenced by our unprecedented destruction of non-human life through our overuse of plastics. And since we seem to be more comfortable accepting animals and plants as fellow guests on the planet—than things like pebbles and garlic presses—I'm going to focus my remarks on such so-called "inanimate" things, because a shift in our understanding of the nature of things can mitigate our environmental predicament; not to mention, greatly enrich our experience.

But before we consider things of the apparently inanimate kind, we should recall that philosophers haven't always understood them as configurations of lifeless matter. Pre-Socratic thinkers like Thales regarded the entire cosmos as being infused with soul, while Plato and the Neoplatonists talk of the World Soul. Ideas were eclipsed by the advent of Christian philosophy, which insisted on God the Creator as the prime animator of things. In the 17th century, the Cartesian thinkers hardened the line between animate and inanimate by denying soul to anything other than the human being, relegating everything else to the category of *res extensa*, or extended stuff. This encouraged the view of the world as dead molecules in motion, as matter, (and in Isaac Newton's words) as inanimate.

But, if we look to the East Asian philosophical traditions, we find a far more congenial understanding of what we call inanimate things, and one that can surely alleviate our environmental predicament. The Chinese thinkers understood the world as a field of *qi* energies. The great Daoist philosopher Zhuangzi talks of the one *qi* that is the world, saying that all beings take shape between heaven and Earth and receive *qi* energy from the *yin* and *yang*. And since it's all one energy, *qi* is not just life energy. It also configures rivers and rocks, as well as the animal and vegetable realms.

Later Mahayana schools of Buddhism regard human beings as already by nature enlightened, which leads to the idea of what they call "Buddha-nature." But then, the Tiantai Buddhist thinkers in China began to question the customary distinction between human beings as possessors of Buddha-nature and the rest of the myriad things; and, to question likewise the distinction between sentient beings and non-sentient. The Tiantai patriarch Zhanran argued in the 8th century that not only living beings, such as plants and trees, have Buddha-nature, but also pebbles and even specks of dust. On the basis of the illusory nature of all conventional dualisms, he argued that the distinction between animate and inanimate is completely arbitrary. Buddhist thinkers in Japan took this idea further. Brimming with Buddha-nature, all things are able to express the Buddhist understanding of how the world works. In the 9th century, Shingon Buddhist thinker Kūkai argued that all things expound the Dharma and explicate Buddha's teachings. They expound not only through sound, as if preaching the Dharma, but also through images that can be read like Buddhist sutras, or scriptures.

Four centuries later, the Zen master Dōgen extended this idea to include explicitly human-made things, such as pillars and lanterns. He says to his monks, "Look to trees and rocks, fields and villages to expound the Dharma. Ask pillars about Dharma and investigate with walls." They can attain the Way, he tells them, through wholehearted interaction with such things. This is because grass, trees, tiles, and walls practice together with you. They have the same nature, the same mind and life, the same body and capacity as you have. But certain activities are more relevant, right, because more central to our existence than others: eating, for instance, and clothing ourselves, and certain things likewise. Eating bowls also used in begging for food and Buddhist robes are both emblematic of the Zen monk and the transmission of the Dharma down the generations. In some cases, particular bowls and robes were actually handed down along with the transmission of the teachings. Otherwise, they were replaced along the way. "Experience eating bowls," Dōgen says, "as belonging to the Buddha ancestors, or as actually the body and mind of Buddha ancestors. Even if the bowls we use haven't been handed down from a previous practitioner, we are to treat them with respect, as exemplars of practices that endure from generation to generation. Through using them, we gain a sense for the practices of others from other places and times." Although they often outlast their human makers or users, eating bowls are not limited to new or old, ancient or present. "Eating bowls," says Dōgen in his most comprehensive characterization, "are a compound of all things." The Buddhist robe is traditionally an assemblage, patches of fabric sewn together, optimally soiled in discarded cloth that has been washed, cleaned, and made pure. Dōgen says not only that a kāṣāya is the Buddha body, the Buddha mind, but even that it's the essence and form of all Buddhas. The Buddhist robe and eating bowl may be extraordinary things, but they are emblematic of how ordinary things can be if we adopt the appropriate viewpoint. Dōgen recommends to the monks who work in the monastery kitchen that they need to stop seeing with ordinary eyes and thinking with ordinary mind, so that they can engage more fully with the magical process of cooking. They should use polite speech in Japanese when referring to the materials of their craft. Dōgen says, "Use honorific forms for describing how to handle rice, vegetables, salt, and soy sauce. Do not use plain language for this." By paying close attention to the utensils and pans, they will know where in the kitchen they belong. But, the order of the well-ordered kitchen doesn't derive from the plan in the head of the cook who's in charge, but rather from his paying close attention to suitabilities suggested by the things themselves. When it comes to doing the cooking, Dōgen calls the creative handling of utensils and ingredients "turning things while being turned by things." We need a sense both for how things are turning out, so that we can align ourselves aright, and for how our turning is, in turn, affecting what is going on.

Now, some people might regard the Zen practice of befriending things in this way—by being sensitive to their inclinations—as a piece of naive animism, in which human traits are being projected onto the non-human world. Yet, these ideas are anything but naive. Dōgen is one of the profoundest thinkers in the history of philosophy whose ideas developed out of a highly sophisticated tradition of Buddhist thought. But, more importantly, this sense of animism—soul projected onto material things by simple minds—refers to a remarkably recent and parochial notion. The English cultural anthropologist E.B. Tyler coined the pejorative senses of the term only 150 years ago. But, the larger point here is this: it's only when you subscribe to a Cartesian mind–matter dualism that you even need a word like "animism." For most people during most of human history, the world has naturally presented itself as animated or ensouled from the start. Animism is a modern and typically condescending presumption.

Marion Hourdequin: Today I'd like to talk briefly about Confucian ethics and the environment and highlight the kind of relational and intergenerational dimensions of early Confucianism that may be relevant to environmental philosophy. Some of the most important environmental challenges that we face today are intergenerational in nature. These include climate change, plastic pollution, nuclear waste, and the long-term ramifications of war and armed conflict. These challenges are also arguably a matter of the character and quality of relationships we have with one another and with the broader world. Confucianism is a tradition that centers relations and also intergenerational connections, and so this tradition may have something important to say about environmental and intergenerational ethics.

In much of Western moral philosophy, I would argue intergenerational ethics is under-theorized, and in analytic philosophy in particular, much of the discussion has focused on whether it's possible to have obligations to future people at all; and also on the non-identity problem, which calls into question whether distant future generations or distant future people can be made worse off by the actions of those living today. Early Confucianism conceptualizes intergenerational ethics differently. Instead of starting with (sort of) fundamental questions that are skeptical about the possibility of intergenerational relations, early Confucianism starts with the assumption of relational embeddedness in intergenerational community. And from that perspective, it offers a basis for understanding the nature and grounds of intergenerational obligations.

I'd like to focus on two questions. One is this: what might an (early) Confucian approach to intergenerational ethics look like? And the second is: how might a Confucian approach to intergenerational ethics be relevant today, especially in (re)thinking relationships with one another and with the broader world? So, to give you a quick preview of my answers to these questions, I'll argue that an early Confucian approach to intergenerational ethics can be grounded in a relational perspective and a relationally oriented ethics. In this perspective, persons are understood as located in a web of social relationships that extend diachronically through time (and this locatedness is fundamental to moral personhood). And what's more: moral development—becoming a fully humane person, or a fully morally developed person—involves cultivating capacities, skills, and habits that generate and sustain mutual flourishing over time. And I believe this approach may be relevant today insofar as it emphasizes a relational rather than an individualistic conception of persons, especially because I think this individualistic conception is responsible, in part, for a number of the tangles that we face in addressing collective action problems like climate change. And this approach also foregrounds a kind of social and intergenerational reason, as opposed to a kind of individual self-interested rationality.

I'm not going to talk much about the details of early Confucianism, but I did want to mention that there are a number of key texts in this tradition: *The Analects* of Confucius, the *Mengzi*, and the *Xunzi*. I'll be focusing primarily today on *The Analects*. All of these texts are over 2,000 years old; the thinkers who developed them lived more than 2,000 years ago, but the ideas remain relevant and salient today and are part of many, many traditions in East Asia and throughout the world. One way in to thinking about Confucian intergenerational ethics is through a closer look at a particular intergenerational relationship in the text of *The Analects*. And this relationship involves Confucius, or Kongzi, and his follower or student, Yan Hui. Yan Hui is a student that is admired and deeply respected by Kongzi, and he has a special place in *The Analects* and in Kongzi's life. I think one of the interesting features of this relationship between Yan Hui and Kongzi is that although Kongzi is older and serves as Yan Hui's teacher, the learning between Yan Hui and Kongzi is clearly bi-directional. Kongzi repeatedly comments on Yan Hui's insights and on how he, Kongzi, or Confucius, learns from his student, not just the other way around. Unfortunately, Yan Hui dies young and Kongzi is devastated. He says, "Heaven has bereft me." Kongzi is so upset that his other students comment on his showing undue sorrow in response to Yan Hui's death, to which Kongzi replies, "Am I? Yet, if not for him, for whom should I show undue sorrow?" I think a critical question this poses is: how might we understand the loss of Yan Hui and the meaning of this loss for Confucius? So, what is lost when Yan Hui dies and his relationship with Kongzi severed? And I think there are a number of possible ways of answering this question. We might think about the loss of potential, the (sort of) curtailment and foreshortening of Yan Hui's life. He dies when he's young, and presumably he and others expected him to live much longer and contribute and engage in many ways over the course of his life to broader society. There's also a loss of a friend for Kongzi. We might also think about the way in which the loss of Yan Hui is the loss of a potential opportunity for a kind of personal immortality for Kongzi, where his projects are carried on by those who come after him. And a sense of loss of continuity in Kongzi's personal values and projects.

But, I actually think that especially with respect to these last two possible explanations, they overemphasize a kind of personal loss that I think maybe isn't central for Kongzi. Although there is a personal loss for Kongzi and the loss of this relationship, I think there's another kind of loss that's not just about Kongzi and his future or his immortality. What's important, I think here, is that Kongzi and Yan Hui's identities are bound up with one another, and that this is an important source of meaning for Kongzi. So, that's something that has to do with a personal dimension of the loss, but there's also a loss of shared engagement in a kind of ongoing intergenerational human project that both Kongzi and Yan Hui were committed to. And this is tied into a broader conception of persons as relational and a relationally based conception of mutual human flourishing that I'll say a little bit more about. So, one of the ideas in early Confucianism that I think is important in thinking about the relevance of Confucianism to intergenerational environmental ethics is this concept of the relational self. In early Confucianism persons are understood as persons in relation, in networks that extend across space and time, and individual identity and well-being are bound up with others, both synchronically and diachronically. This is the kind of conception of the relational self that figures in early Confucianism where others actually enter into and shape our identities in fundamental ways.

Two other ideas that are really important, I think, in kind of understanding the conception of intergenerational ethics that we might construct out of early Confucianism are gratitude and reciprocity. And gratitude in early Confucianism is importantly tied to family relations and to gratitude for the care that we gain from parents or other caregivers, and also from teachers and mentors. And this gratitude gives rise to a sense of reciprocity, but neither gratitude nor reciprocity is understood in a kind of bilateral sense where our gratitude for things that we are given must be repaid directly toward those from whom we've received them. Instead, gratitude and reciprocity can be understood in an extended way, more like a web, a net, or a chain extended through time where we might express gratitude to what we've... for what we've gained from our caregivers to those who come after us, to younger generations.

On this view, there are three elements of a Confucian intergenerational ethic: a relational view of identity and the human good; gratitude toward those who support and shape the development of our identities; and a shared commitment to flourishing human communities that are sustained through time. And gratitude can feed into and support this shared commitment insofar as it's not expressed through a kind of narrow reciprocity, but through this broader reciprocity that I discussed. The Confucian relational conception of persons together with a Confucian conception of gratitude and asymmetrical reciprocity support this shared intergenerational commitment to build and sustain flourishing in harmonious communities. And although early Confucianism doesn't really provide a systematic understanding of this project as encompassing the broader natural world, I think the view might be extended in this way.

How might an early Confucian approach to intergenerational ethics be relevant today, especially in relation to environmental and climate ethics? And I think here we can think about how it's not only human beings and human relationships that shape and support persons' development and their identities, but broader ecological systems also are involved in shaping and making possible meaningful human lives. So if Confucian gratitude is an affirmation and a recognition of important contributions to one's life and identity, it may be extended to broader environments and to relationships with diverse living beings and ecological systems. Lastly, I think this approach contributes to a kind of social and intergenerational reason that provides a contrast to the kind of individualism that's dominant in many contemporary societies, which isolates people both synchronically and diachronically, and alienates people from one another and from the ecological systems in which we are embedded. A relational conception of persons has the potential to re-embed persons and communities and in moral ecologies, and it can ground a different mode of reasoning other than dominant conceptions of individual rationality.

Leah Kalmanson: I'm starting here today with a framing quote from this theorist, Freya Matthews. And she says, "To accept that the world in which we dwell has not only a physical but a psychoactive dimension, that it is meaningful and mindful presence rather than mere brute materiality, is to wake up to the real significance of the ground beneath our feet. Were we collectively to grasp this significance, an entirely new chapter in our relationship with reality would surely be initiated… Industry, engineering, agriculture, all our modes of praxis and production, would be recognized as conditions of encounter and communication rather than merely instruments of extraction. Our entire economy would, in other words, become a vehicle, indeed our principal vehicle, for meaningful exchange with a responsive world." And that's from an article she wrote for *Sophia* in 2010. So, she calls panpsychism here a metaphysical insight, and part of my goal today is to think through the philosophy that does ground that insight in ways that do deviate somewhat as you'll see from the direction that Matthews herself takes, but which I hope remain complementary to her project. She's part of a larger group of theorists who are exploring panpsychism, animism, and questions of disenchantment and re-enchantment as these are related to issues of environmental thought and activism. I'm interested in the ways that indigenous worldviews are often engaged in this context as alternatives to the so-called disenchanted worldview. And my concern, though, like I said, is how we articulate the fuller philosophical picture that grounds that engagement. My goal here is to develop some language and philosophy of religion that is useful to this larger environmental project.

So let me start with the problem, put bluntly, which is this: calling a tradition polytheistic is not a complement in the philosophy of religion, right, whereas belief in a single, transcendent God has been considered rationally defensible. Belief in anything else is usually dismissed as ungrounded superstition, right? This has theological roots but political implications as well. There's a 2005 book which I love, *The Invention of World Religions*, by Tomoko Masuzawa. And this gives us some sense of the politics. As she says in that book, "Take a look at any textbook in world religions. Why, after the first ten or so chapters on the usual suspects, do we always have some catch-all chapter under a heading like 'Native' or 'Indigenous,' sometimes 'Primal' or 'Primitive'? Why are regions as diverse as the Americas, Oceania, all of Africa classified together in this single chapter?" As Masuzawa says, "This chapter is a descendant of 19th-century European scholarship and its hierarchical mapping of so-called 'world historical, great civilizations' and others that are" as she says, "perhaps not so great." The latter (quote) "used to be uniformly called "primitive religions" in the earlier days, but more recently have been variously termed "primal," "preliterate," "tribal," ... The restless shifting of appellations may be a measure of the discomfort felt by contemporary scholars of religion in their effort not to appear condescending to those people who used to be referred to as savages." My primary concern is that philosophy of religion has some response here.

Let me offer a concrete and personal example. In 2003, I relocated from the crucible of the racially divided South where I grew up to the vividly neo-colonial reality of present-day Honolulu (I was attending the graduate program in philosophy at UH Mānoa). My example here will concern legal challenges over agricultural patents from around this time, and it's precisely this kind of issue that I think philosophy of religion should have a response to. In the Hawaiian tradition, genealogy connects the indigenous people to the land. The generative forces of sky-father Wākea and earth-mother Papa birth the islands of Hawai'i and Maui as well as a human child—a daughter—Ho'ohokukalani. Wākea and Ho'ohokukalani then procreated in secret. Their first child, Hāloa-naka, was stillborn. From his buried body grew the first kalo plant, that is, taro root. The second child was a human boy, named Hāloa after his deceased brother, and he's the ancestor of all later Hawaiians. A kalo plant, as older brother of the people, remains central in Hawaiian thought and practice. In recent decades, so-called "modern" farming methods moved away from traditional practices that once preserved the genetic variety of kalo. This resulted in plants vulnerable to a blight. Researchers at UH Mānoa produced genetically modified resistant strains and patented them in response to a 2002 legal challenge over these patents. Local farmers rejected the university's right to ownership because, as this person is quoted here, "(f)rom a Hawaiian perspective, any kalo is our ancestor." In the statement intended to acquiesce to these concerns, the university's Vice Chancellor Gary K. Ostrander said, "Taro is unique to the Hawaiian people in that it represents the embodiment of their sacred ancestor. As such, it is appropriate to make an exception to our standard policy of holding all patents." So, it represents the embodiment of the sacred, right? So, Ostrander portrays kalo as the representation of the sacred.

But, as philosopher Julia Morgan makes clear in her study of these legal disputes, the Kanaka Maoli did not oppose the patent because the kalo symbolizes a Hawaiian ancestor, or its metaphor for Hawaiian heritage, or any other argument that downplays the agency of the kalo itself. As she says here, "Simply, 'Ōiwi Maoli opposed the patents because a relative cannot be owned." There's a mismatch of worldviews here, right, and what is at stake. Rendering Hāloa as one representation of the sacred is an attempt to draw Hawaiian tradition into the fold of world religions, right? But Hāloa is not one representation of the sacred on this buffet table of the world's great traditions. Hāloa is a local God. Does philosophy of religion have language for this? Generally, I think we can see monotheistic theologies do not, and it was in search of better language that I first got drawn into the literature on re-enchantment and panpsychism, which I thought might give me some new approach to local gods, especially as these have often had a connection to or are identical with natural elements, geographical features.

I was trying to think about this living landscape that I first read Freya Matthews's 2003 book *For the Love of Matter: A Contemporary Panpsychism*. In it, she takes the generally non-dualist model as underlying the metaphysics of this worldview. As she says here (quote), "the universe may be conceived as a unified, though internally differentiated and dynamic, expanding plenum"; it's "necessarily self-actualizing." She calls it a "cosmic self" or a field of subjectivity that is "logically prior" (in her words) to individuated entities, much like, as she says, "primary qualities" are logically prior to "secondary ones." She explains, "For under its extensional aspects, the universe is differentiated into local modes, some of which may be capable of experiencing themselves as relatively distinct unities, or centers of subjectivity, separated out from the greater whole," but these "secondary subjects" can be viewed as "distinct individuals, even while it is understood that such individuals are ultimately continuous with the primal, indivisible whole." She further explains the self-interest of secondary subjects on a model of ecological balance, claiming, "True, inter-species relations are typically, in undisturbed ecosystems, mutually sustaining at the level of populations... Predation and parasitism are thus, in intact ecosystems, generally finely tuned forms of environmental 'management.'" She extends this thesis in comparative terms: "Such a happy

prearranged, basically ecological harmony between the desires of the Many and those of the One may be taken to constitute what might be called, after the Chinese notion of the
Tao, the 'Way' of the One and the Many."

Note here that Matthews's vision of ecological harmony is
what we could call "depoliticized." All power balances are
only apparent, right; that between predator and prey, parasite and host. From the larger perspective of the cosmic self,
these are all in fact harmonious. I want to work with her
insights about the psychoactive aspects of the biosphere, but
I want to articulate the underlying philosophical picture differently. To the extent that I worry about drawing Hawaiian
tradition into the fold of world religions, I also worry about
drawing Hāloa into this happy, prearranged harmony, where
he's only but one avatar of the greater cosmic self. Because
Hāloa has a specific family and it's not my family; and I
think that's important to say. When I lived in Hawai'i, I was
at best a guest in Hāloa's house. I could claim no kinship by
any spiritual approach. My relationship with Hāloa was political. These political relations between mainlanders, non-
indigenous state residents, Kanaka Maoli, and local Gods
mark not only the legal battle over kalo patents from a few
years ago, but ongoing Hawaiian activism against further
telescope construction on Mauna Kea. These are live issues.
If we're seeking a re-enchanted worldview that takes Hāloa
seriously as a local God, then it has to retain the capacity to
address these political dynamics.

In what I hope is a contribution, I take my cue here from
going back even earlier in the Hawaiian genealogical narrative. This is the start of the classic account. Here translated
by Queen Lili'uokalani during her house arrest prior to her
forced abdication to American rule, as you can see, this is a
story of spontaneous emergence. From heat, from darkness,
from power comes forth a slime, and if you continue reading
the narrative beyond this part, from the slime comes forth all
the first things that grow in the world. So, this image brings
me back around to material that's closer to my own research
areas in Chinese philosophies. In general, Chinese cosmology shares these patterns of generation and spontaneous
emergence seen in the Hawaiian model.

Both Daoist and Confucian cosmogonic narratives follow this basic pattern. First, there comes a primal state. It's
described variously as formless, undifferentiated, chaotic, or

empty. Next, there spontaneously emerges the initial distinction into the polar forces of *yin* and *yang*. And then, from the increasingly complex interactions of *yin* and *yang* arise the myriad things of the cosmos as we know it. All of these models accommodate Matthews's point that the stuff of existence is both material and psychoactive. But I note here that it would be awkward to describe *yin* and *yang* as manifestations of the primordial *qi*, or manifestations of this primal source. Some strands of Chinese thought portray the emergence of polarity as a transition from chaos to structure, others as the disturbance of a once pristine tranquility. But, either way, it's not the manifestation of some generic primal stuff into two specific forms or modes. We can't capture the relation of the primordial source to *yin* and *yang* through either a logic of genus and species or of primary and secondary qualities.

So here's where I think the Hawaiian model, the Chinese model, and others give us a different philosophical picture than the happy, prearranged harmony that Matthews associates with the notion of Dao. And it's a picture that retains the political dimension I signaled earlier. The cosmogonic position of spontaneous emergence offers, I think, an interconnected, non-totalizing explanatory framework that allows for meaningful differences between local and specific spiritual phenomena to come into philosophical view in ways that models grounded in theological transcendence or mystical holism do not. Models of spontaneous emergence allow for astonishing possibilities—ghosts, gods, spirits—but without invoking anything beyond the same vital, dynamic, primal stuff that gives birth to us all. They can account for certain naturalistic explanatory models without thereby committing us to an entirely mechanistic, deterministic, or otherwise reductive scientific worldview. I think this blurs the distinction between the enchanted and the disenchanted in ways that intervene in the dominant Eurocentric narratives of modernity, which is precisely Matthews's own goal in the quote that I started with. These issues of cosmogonies, right, like many things we've seen here, have philosophical roots and political implications. If we're judging work in philosophy of religion in terms of how it empowers marginalized voices or intervenes in hegemonic Eurocentric practices, especially as these inform environmental policies that seek to take indigenous worldviews seriously, then the cosmogonic narrative of spontaneous emergence will come to the fore as a capacious and inclusive theoretical framework.

David Storey: I will wrap us up on a personal note, but first I'd like to read a quote from one of the articles that Leah sent along that I thought reflected the spirit of the conference very well, as well as much of our discussion today. She wrote, "Perhaps it is no surprise that I ended up in comparative philosophy, trying always to be the best guest possible in house after house." I think that's a beautiful image to think about. House after house, both at the very local level, of neighbors, and at the larger level of cultures. But then even at the level of the Earth itself: that we're guests, even when we're traveling from one house to another, in the larger house that we all share. One reason that I reached out to all of you for this panel is that—in this I'll put my own little editorial spin on this—I became pretty convinced early in graduate school that the future of philosophy was: a) comparative; and b) environmental. Because the future of humanity was: comparative, in the sense of greater intercultural exchange and greater reckoning with our environmental problems. So, I think that it's interesting to draw in these Asian perspectives, because of the way the world is moving—with the rise of China, and increasing globalization in both bad and good ways, I guess you could say—it is going to force us to reckon with these cultural differences and draw on the best of what we have as humanity, East and West, ancient and modern, to deal with the climate and environmental crises. So, I want to thank all of you so much, both for the conversation as well as for your scholarly work.

ECOLOGICAL ENDGAMES
Speakers: Ariel Salleh, Michael Zimmermann, David Storey

David Storey: Michael you were present at the creation of environmental philosophy as a field. Can you talk a little bit about how you got into environmental philosophy, before it was a thing? What drew you to that area of topics?

Michael Zimmermann: In the sixties, the mid-sixties, these environmental issues were starting to come up. In 1962, Rachel Carson published how upset she was by these revelations, about how DDT was affecting the bird population. And then in graduate school, there was nothing about any of this, in undergraduate or graduate school, in the college-academic level. But I maintained an interest in it and then when I was in (I had a Fulbright Fellowship) Leuven in Belgium, 1972 to 1973, that gave me an opportunity to do a lot of reading. I was writing my dissertation and I was also doing a lot of reading. I read most of the two volumes Heidegger wrote about Nietzsche. They came out in the 30s and 40s (that is, he lectured them in the 30s and 40s), and they had just come out in published form in 1970 or 72. It was there that I got his interpretation of the rise of modernity. Heidegger was famous for being in philosophy, but he did not speak about nature very much. In fact, when he did, he did so in problematic ways. But here was a complete revelation to me. So that's how I got launched.

DS: Heidegger's critique of modernity is a technological way of interpreting the world, where he connected that with the very roots of Western metaphysics. So that was sort of your intellectual on-ramp.

MZ: Right. And that took me a while, you know. I wrote my first paper on this in 1976, titled "Heidegger on Technological Nihilism" or something like that, but I kept learning more and more about what he was up to. Then I wrote a book (*Heidegger's Confrontation with Modernity*) and that's where I came up with this phrase "productionist metaphysics." That's what Heidegger is all about in the mid-30s and beyond. And that is to say that most of metaphysics springs from Plato and Aristotle, both of whom use metaphysical distinctions in terms that are reminiscent of handcraft production, you know. Plato's Forms are blueprints, the demiurgos who forms things as the builder. Aristotle's discourse is shot through with reference to making things. The four causes and stuff things are made of. Heidegger says that if you follow this out through the history of philosophy, you end up with, quite surprisingly, Nietzsche.

For productionist metaphysics, to be means to be produced. So God would be the big producer at first, but later on God falls away, and humans become the architects. What it boils down to is that humans reason about it; our capacity to think, our ability to make judgments about what's real and what's not, through Descartes and the rise of mathematical physics, humans get into the driver's seat. We become, in effect, the masters and, in some sense, producers of nature. It becomes an object for us, centering a reality on things around us. It really completes the Kantian turn of claiming that the human mind shapes and organizes nature as it shows up for us. Heidegger takes us to the next step and endows the human with the kind of metaphysical power which Kant had never envisioned.

DS: So part of what I hear you saying is that even the emer-
        gence of nature as a concept, as this allegedly separate
        sphere is a modern construct. It's an intellectual conceit
        that not only does not accurately capture the real world
        but can actually cover it up.

MZ: Well, that's a great distinction. And the big figure here—
        Heidegger makes him the big figure like everyone else—
        is Descartes, who develops Cartesian dualism: what's real
        is either consciousness or extended matter, and they don't
        share any qualities between themselves. They're yoked to-
        gether somehow: mind and body. And, moreover, for him
        mathematical physics—he invented analytic geometry—
        would provide the way in which to interpret the world, in
        the sense that it provided us with the most certain types
        of representations of what things are, why extension is
        the essence of the material world. And if you master your
        comprehension of extension and motion, then you become
        master of the universe, in effect. So, the human subject
        stands over against nature and can transform it in ways
        that he or she chooses based upon our capacities.

DS: And you mentioned a few moments ago the phrase "tech-
        nological nihilism." So there's a connection between this
        particular view of nature that emerges out of Western
        metaphysics and a particular position about value and
        the value of the non-human world. That's where this
        particular critique of Western philosophy feeds into en-
        vironmental concerns, because it's pretty easy to see the
        connection how if value is just a human construct, then
        we have no obligations to anything beyond ourselves.

MZ: Right. There are two issues: of nihilism, and the val-
        ues issue. For Heidegger, nihilism doesn't have to do
        with a decline of values. He would say it involves the
        ascent of values, because, for him, value—taking off
        from Nietzsche—is a perspective that provides a way
        of enhancing power. That is an unusual sense of value,
        but that's what Heidegger says Nietzsche says, and,
        probably, there's a lot to what Heidegger's reading has
        to say regarding Nietzsche's world of power. But, for
        Heidegger, the *nihil* has to do with the emptying-out
        of the substance of things, the reality of things, their
        own-ness, by our turning them into objects. We cancel
        them out, we don't let them reveal themselves in their
        complexity. However that's to be understood, so much
        is omitted. Now, tremendous power and control result

from that, so let's not forget that the heart of modernity is electrification basically, right? Nevertheless, so much is left out; it's like the world is emptied of meaningful content beyond mathematical physics.

DS: Ariel, you helped develop the field of eco-feminism as an alternative to deep ecology. So maybe we can just start with some general definitions of what those two different positions are, how they differ, and so on. Maybe you could start us off with eco-feminism and then Michael could talk a little about deep ecology, and then, what are the strengths, the weaknesses, the disagreements between them?

Ariel Salleh: Well, first of all, I think eco-feminism is a grassroots movement. It really emerged spontaneously from women's activist experience against toxic neighborhoods and nuclear testing on indigenous land. Because women's reproductive labor, women's role in society which was given to women in ancient times because women were seen as closer to nature than men, although, of course, all humans are nature in embodied form.

The deep ecologists didn't know the women part, but they knew the link... they were reaching forward to mend the link between humanity versus nature, and there, I totally agree. The first statement I think I made in the "critique of deep ecology," the first go of making that critique against Arne Næss and Bill Devall in 1984, was that the claims of the deep ecologists laid about by Næss and Devall were indisputable about the need to rediscover ourselves in nature and so forth. They were indisputable. I think the deep ecology team was a little bit destabilized by this feminist critique coming out of the other side of the planet that they disregarded that fundamental agreement.

To me, it seemed that what was necessary to reconstitute the humanity–nature link was to look at the labor that women do, which is called reproductive labor, as distinct from productive labor in the mainstream capitalist economy and so forth, to examine these processes in a phenomenological sense. How is this physical, material bridging of human cycles and natural cycles being carried out?

Western civilization, industrial civilization is founded on an ancient violence splitting humanity off from nature. But meta-industrial labors—this is my term which I invented in critiques with eco-socialists— at the margins of capitalism practice a synergistic epistemology, a synergistic epistemology which balances and restores the metabolism of human and nature as one systemic flow. This work embodies a sound ecological ethic. What if the suggestion was made that, in building a democratic solution to the global ecological crisis, it's precisely these marginals—housewives, peasants, indigenes—who can show the way under the global economic system? Common land water, biodiversity, labor, and loving relationships are pulled away from an ecologically sustainable and culturally autonomous web of self-sufficiency. By the north versus south, one-zero logic, black-white, man-woman, humanity-nature dualities, modernization turns people into human capital and lands into natural capital in order to benefit an international minority class of entrepreneurs and government hangers-on. For centuries, communities from Africa to Oceania to South America and beyond have struggled against this appropriation of their livelihoods and have established academic disciplines, I'm afraid, that tend to support that colonizing mindset, despite its devastating effects. Likewise, mainstream liberal feminists and left activists often favor emancipation via the industrial paradigm, assuming that an equitable redistribution of the social product is possible. Neither standpoint makes any thermodynamic sense. As far as the protection of life-giving metabolic value goes, this was something I was hoping we'd talk about: the deep ecologist's notion of intrinsic value, because my materialist, my metabolic, embodied materialist concept is metabolic value—and I also use that to contest the eco-socialists who are on about use value and exchange value which are intrinsic to capitalism.

Consider the account of humanity-nature metabolism in eco-feminist Vandana Shiva's case study of local Indian forest dwellers. She says that it's in managing the integrity of ecological cycles in forestry and agriculture that women's reproductivity has been most developed and evolved. Women transfer fertility from the forest to the field and to animals. They transfer animal waste as fertilizer for crops and crop by-products to animals as fodder. So, this partnership between women's work and nature's work ensures the sustainability of sustenance. Here's a statement of humanity-nature co-evolution, a regenerative economy fully compatible with scientific complexity. This livelihood practice rarely takes more matter and energy than needed.

Turning to meta-industrial labor in urban economies, the German ecology activist Ulla Terlinden spells out the phenomenology of household reproductive labor that is also at the periphery of capitalism, because households are not considered economic, although, in fact, this work subsidizes the capitalist productive system. Terlinden writes that the household requires of women or men (if they're doing it) a broad range of knowledge and ability. The nature of the work itself determines its organization. The work at hand must be dealt with in its entirety as a system. The work must possess a high degree of personal synthesis, initiative, intuition, and flexibility. Contrast this sensitive empirical engagement with the fragmented industrial division of labor, from the shattered mindset of the investor through to the numb assembly line operative.

In the context of parental skills, US feminist philosopher—this is way back in the 1980s—Sara Ruddick introduces the word "holding" in order to mark parallels between good parenting and good ecological reasoning. She says that to hold means to minimize risk and to reconcile differences rather than sharply accentuate them. Holding, by a man or a woman, is a way of seeing with an eye towards maintaining the minimal harmony, material resources, and skills necessary for sustaining a child in safety. It's the attitude elicited by world protection, world preservation, world repair. While minimizing risk, often in the face of dire material uncertainty, holding is the ultimate expression of adaptability, and while science, as usual, is marred by the positive separation of fact and value, space and time, a cautious awareness of interconnection is common sense in this embodied materialism.

To realize eco-sufficiency with global justice, the most effective millennial goal—or sustainable development goal for that matter now—will be to give back land taken away from peoples in the name of development. This presents a major challenge in an era when land grabs for mining, water extractivism, and other corporate projects impact every continent.

DS: Michael and I were talking about how he got into environmental philosophy, and part of that was through Heidegger's critique of Western philosophy, which he described as a "productionist metaphysics." It was very much this idea of (often represented as the male sort of creator or producer) bringing things into being, sort of out of nothing. And I think it's pretty easy to see how that has a very masculine connotation that plays down or ignores the feminine dimension of reproduction that you're talking about. But you referenced a number of things from deep ecology as well. So Michael, maybe you could zoom out a little and just lay out some of those basic ideas from deep ecology or anything you wanted to add to Ariel's.

MZ: Well, first, Ariel, you did terrific things during the 80s and 90s, and I learned a ton from you and other people involved in eco-feminism. And so I was one of those deep ecologists who wanted to find a way to be in conversation with what you were talking about. I have to say this: I don't know if you've ever seen this BBC show "Call the Midwife." Do you know this show?

AS: "Call the Midwife"?

MZ: Yes, it's fantastically popular. Now, these are midwives in the east end of London after World War II, late 50s, early 60s; the poorest part of London. And they are part of an Anglican religious organization. Some nurses are secular, other midwives are Episcopalian, Anglican. But, anyway, the point is the depiction of labor. These are stories about people in extreme poverty and are subject to violence and are struggling. The delivery scenes are so powerful at times that you can't help but being moved as they are by the bringing forth, the manifesting of this new life. It is and was a constant reminder of what it's like for part of our population, half our population. Now, not every woman has a child, but those women that do and have several: I mean, this is it! We're not going to be here without that, and the terror, the pain, the uncertainty, the death, and harm. I mean, all those things come into play.

AS: I'm so pleased you said that because I felt… for me, I think, becoming a mother was… a major learning for me was that it is a shock, it is a terror. The pain is indescribable, the uncertainty. And, of course, the whole Western civilization has protected men from knowing that. There's enormous taboo. I notice even mothers don't talk to their children about that very central, formative experience. The moment that you i.e., the child came into this world is not discussed. Everything is hush-hush; it's put away in specialist medical centers. It's as if men are protected from this knowledge of their origin, because to acknowledge that would be to give women an enormous power I think.

Of course, this goes back hundreds of thousands of years: the emergence of a patriarchal culture carries forward this dissociation-from-life process, this coming right forward to Western civilization and the Enlightenment and the rise of science and so on. It's very marked in our contemporary culture, the dissociation. I call it the "one/zero" mindset, which means that everything is either this or that, subject or object... black or white— white or black, actually, with the first term always the valued term and the second term the object.

MZ: So, you see that dissociation in our treatment of the natural world, and that's the entry for eco-feminism.

AS: Exactly. As I became more involved in activism, and over the last 20, 30 years, you realize that the planet, it's all interlinked and it's all expressed and bound and entangled with capitalism and colonization.

DS: You also mentioned a midwife philosopher, which I thought was interesting because, classically, Socrates's task is understood as a philosophical midwife, but it's in intellectual terms, right? So, there's this disconnect, there's a sort of abstraction of the actual concept. And Michael and I were talking earlier about how very much the culture of modernity leads to this "materialism." And I was interested to see that you described your work as "embodied materialism," which might sound like a redundant phrase, but I think I see there's a deeper dimension of what you're talking about. The disembodied materialism that predominates in our capitalist, consumer culture is actually an abstraction, and it pulls us away from the brute materiality of our being. So, those are just a couple of things that I wanted to pull out from what you were saying that sounded really provocative if I'm getting you right.

AS: To clarify the use of the word "materialism": because the capitalist sense, our every sense of "Oh, she's materialistic, she wants a new dress every week and so forth, a new refrigerator and what not"—that's actually not the use of materialism that I mean; that's the everyday use. It's not the use of materialism I have mentioned, which is the philosophical one. And my urge to look at or consider an embodied materialism was actually part of a debate I was having with eco-socialist thinkers like James O'Connor and John Bellamy Foster.

DS: Michael, can I pause you there, actually. I think this would be really interesting for listeners to hear because I don't think it's something that a lot of people know,

you know. When we think of the environmental movement, we think of the 60s and 70s and Earth Day and the
formation of the EPA and all this environmental legislation. But, there was an environmental movement in Nazi
Germany. So, maybe you can talk a little about that and
how Heidegger was, in that orbit.

MZ: Beginning in the early 20th century, there became a
youth movement with young people tramping around
the countryside, sort of reminiscent of our 60s in a certain way, and it led… that was one of the many manifestations of incipient romanticism regarding nature. But,
after World War I, part of the rise of Hitler's National
Socialism had to do with a wing of his party emphasiz-
ing the connection between blood and land. Well, actually, it was pretty important to the part. So, their motto
was… the Nazi motto, one of the big ones, was *Blut und
Boden*, "blood and land." That is to say, pure blood and
pure land. That was the supposed premise, as it were, of
that statement. And so, the whole idea of purifying racially had to do with creating healthy blood in Germany
and that meant the exclusion of Jews and Gypsies and
ultimately the creation of extermination camps and the
whole horrible history of National Socialism.

Now, Heidegger embraced Hitler and, years later he
wrote in notebooks (that have been revealed in the last
five or ten years) in which he has a few explicit anti-
Semitic remarks. Very troubling, although you do not
find these in any of his published works. But, he did
speak in favor of Hitler and separated in the mid-30s,
because Hitler just was not up to the task assigned for
him, which was to be a true revolutionary new beginning of Western history. So, that just didn't happen.
But, part of Heidegger's attitude was shared by many
people in Germany. In fact, I remember coming across
this book which someone referred to me. It was a book
in German; it came out in the 1930s, in 1936 or 1937
when nature protection had become a huge thing in Nazi
Germany. And so, you see a nature protection magazine
with photographs (this is 1933, 1934, 1935) of a Nazi
Youth marching around in the countryside with their
swastika flags because of the close connection between
protecting nature, protecting the *Heimat*, the "home-
land," nature, versus outsiders and protecting the people. I mean, the environmental movement went way
out with the Nazis in many respects, because of their

revulsion, shared revulsion toward Communists, Marxists, which they thought were antithetical to nature.

The proximity between Hitler's movement and the incipient green movement in Nazi Germany was sufficiently well-known that you could not get an environmental movement going in Germany until the 1970s because there was such pushback to that; the issue was about how protecting nature could be done without protecting Nazi pure blood. So, finally, the Green party broke through, but it took a long time for that. And I was reading all this and I got this realization that I have to be careful with promoting Heidegger as a forerunner of deep ecology. But I will say this: the loss of the possibility of being a conservative in a traditional sense, I mean people who want to preserve, including preserving their land, was vital because you depended on it. That was your source of production.

AS: Right. I mean, how modern this is now, where the whole environmental movement is moving toward de-growth and food sovereignty, permaculture. And my own analyses of reproductive labor, which are not only women's sexual reproductive labor. But, the meaning of reproductive labor, of course, talks about instilling social mores and the reproduction of cultural mores and economic reproduction and the reproduction of habitat—which we would probably... I don't know whether we want to get onto that—but, the work of Vandana Shiva in India, herself a scientist in the beginning, has the most exquisite unpackings of peasant labor and indigenous labor in terms of its relationship with nature. It inspired me to no end.

MZ: Well that brings, I think, up a point that is very pertinent to our contemporary moment: What you were saying about the proximity of native and indigenous people to nature. I mean, the stuff they need for life is much closer, much less mediated, right? They don't buy stuff in plastic at a grocery store. I mean, the whole thing is different in that respect. But, think about the implications of the cell phone. I mean, there was a book that came out about twenty years ago called *Last Child in the Woods*. What a great title, right? Who goes into the woods? You've got your iPhone, right? It's a big transformation that is going on right now. The intensity of technological development, the exponential growth of artificial intelligence and virtual reality.

Part V

Ecologies of the Earth

Chapter 12

Environmental Epidemiology

Peter Klapes in conversation with
Kate Burrows

Peter Klapes: "Hosting Earth" is grounded in the belief that we are both hosts and guests of the earth. Can you tell us a bit about your work in terms of this concept of ecological hospitality?

Kate Burrows: I am an environmental epidemiologist by training. I study public health sciences and I am broadly interested in the ways in which the environment influences human health. I am interested in studying the how rising temperatures, air pollution, and extreme disasters (hurricanes, landslides, fires) relate to and affect human health. In my own work, I use both qualitative and quantitative methods to study how the environment impacts population health and wellbeing. This question is fascinating as it relates to hospitality—especially since we're often not able, in the field of epidemiology, to consider such theoretical underpinnings. My work in epidemiology has actually focused mainly on the *inhospitability* of environments to human habitation. For example, in some of my work, I have been able to quantify the increase of hospitalizations as a result of tropical cyclones in the United States and in a new project we are looking at extreme heat effects and how that impacts mental health. Such changes in the environment are making it difficult for people to continue to live in places they have inhabited for many, many years.

Another point that I have observed in my research is the vulnerability associated with hosting and being hosted. So for example, we have found that those who are more reliant on the land—like farmers—often are most at-risk when faced with environmental disasters. This vulnerability has certainly always been part of the human condition and part of living on the earth, but where it is becoming a larger problem is in relation to ecological disasters. As we consider the reciprocity of hosting, I think these problems do indeed stem from our failure to adequately "host" the earth. Much of my work is certainly pessimistic, but in my work I have begun to attempt identifying ways in which the environment can

DOI: 10.4324/9781003456940-17

be beneficial and provide a place of refuge, especially for people who are displaced and migrating. The parks and trees around us can perhaps help people heal from disaster, so we get a healing benefit from that which is the epicenter of so many disasters and catastrophic events.

PK: There seem to be ways out there—and I am hopeful, especially given what you've said about the benefits of nature itself—to solve these problems and that very pessimism you've identified. In solving a problem, though, it is often helpful to consider the root cause of a problem. What do you posit might be the causes of the problems you study?

KB: Well, the way that the environment, for example, impacts human health varies quite a bit relative to the type of exposure that we are talking about. I am interested in a number of different expo-sures, or events, but each impacts humans in such a different way. For example, in hurricanes, we have seen an increase in hospi-talizations and respiratory disease after hurricane exposure. You would expect to see damage, injury, and those types of health im-pacts, but we're seeing these other more chronic-disease impacts. And what we have hypothesized is that what we are seeing is the result of hurricane-caused power outages. People who are on ventilators, during a hurricane, might experience a power outage and might therefore need to be hospitalized. This sort of research is a way of studying the indirect pathways by which specific environmental exposures might impact population health. But ultimately, the root cause will vary a great deal depending on the specific environmental exposure and the specific context in which it is occurring.

PK: So what are some possible solutions? What is one concrete thing we can do to deal with this—to help mitigate some of these nega-tive impacts of the environment on people?

KB: My work really focuses on adaptation. I think one thing we can do is identify the specific causal pathways that are impacting popula-tion health. This is the best way that we can intervene to protect peoples' health in the long-run. For instance, if we are thinking about the impacts of extreme heat, especially on the homeless population, we can identify perhaps where people can benefit most from cooling centers and such solutions. Research that tar-gets specific populations and that incorporates geographic and spatial data are always most effective.

PK: You mention cooling centers. And this piques my interest because from what I understand, artificial cooling—air conditioning and its related components and chemicals—is so detrimental to the environment. It is a difficult quandary, but how do we balance the need for cooling centers and taking care of each individual person and the fact that some ways of living and indeed some necessities of human life contribute so negatively to the earth and to our ecological habitat and then lead to these issues that you mentioned at the outset. We're in a bit of an endless cycle, no?

KB: That's a great point, and I think that part of that is why tackling mitigation—and not just adaptation—is so important. At the same time that we're considering adaptation, we must be thinking about working towards more sustainable energy sources. You are certainly correct that air conditioners have such a huge trade-off here. Air conditioning requires power and energy and to use them we are burning fossil fuels. But if we focus on specific populations, we can develop targeted interventions and make them more efficient. So maybe we don't need cooling centers across an entire city, but just concentrated in a specific area; or maybe we need cooling centers only above a certain temperature threshold. Making those policy decisions requires evidence, and that's the type of work that I aim to produce.

PK: I know that in your work you also focus on migration. Tell us a bit about that side of your work.

KB: Yes, a lot of my work focuses on the impact of environmental disasters on migration. I am interested in understanding specifically who is at risk of migrating and also then how that movement impacts health outcomes. My dissertation work focused specifically on this topic in Indonesia. I spent a lot of time in Central Java, in a community that was frequently affected by landslides. Despite the fact that landslides occur all over the world and displace millions of people each year, there is not a lot of research on landslide displacement specifically. In the field, we found that there were so many people who were affected by repeated exposures. Landslides are so destructive and it is just so jarring to see just how quickly they can impact a community, usually with little to no warning. Folks often have no time to prepare for a landslide, in the way that they can for a hurricane, for instance. But they are happening much more frequently now, and people are having to continually resettle into their communities. That process of resettlement is difficult of course as landslides are so recurrent these days.

We often hear and think of migration as a process happening on a large scale. People will certainly be displaced, but it will not be all at once, and the migration will most likely happen over short distances. So the alarmism that we have heard about in recent years—related to migration and its security implications, especially—might actually be based on unfounded concerns.

PK: What might be some specific examples of climate migrations to-day that you have seen in the field? Where is it happening specifically? Where are people coming from and going to?

KB: There will certainly be a point in our lifetimes at which we will all know people who have been affected—and have migrated—as a result of these disasters. We are currently seeing climate migration in both coastal areas and in areas affected by drought. We are also seeing a lot of increased risk in some non-industrialized and newly industrialized communities, particularly some small island communities. These are very challenging cases, because when you're talking about a small island, there really are not that many options for refuge, so people will need to move elsewhere, across international borders.

Climate migration also poses a challenging question about requiring people to be hosted by other people. A lot of times, places where migrants can move to are places where people are already living. And there is a challenge in considering how identities can be retained on both sides of the equation, guest and host.

In the face of the complex challenges posed by climate change, it's important to recognize the need for a multi-faceted approach that combines both adaptation and mitigation efforts. My work mainly addresses the former—I try to identify how a changing environment affects public health so as to uncover opportunities for intervention and improving health and wellbeing, for both communities and for individuals. Ultimately I think we have a responsibility to work towards supporting the health of populations around the world who are already being affected by climate-related extremes today.

Bibliography

Burrows, K. & Fussell, E. A life course epidemiology approach to climate extremes and human health. *The Lancet Planetary Health* 6, e549–e550 (2022).

Burrows, K. et al. A systematic review on the effects of long-term, chronic climate change on mental health. *Nature Mental Health* 14, 4026 (2024).

Ebi, K.L. et al. Extreme weather and climate change: population health and health system implications. *Annual Review of Public Health* 42, 293–315 (2021).

McMichael, A.J. & Beaglehole, R. The changing global context of public health. *The Lancet* 356, 495–499 (2000).

Chapter 13

Ecology, Economics, and Ethics

Stefano Zamagni

Environmental ethics, whether ecological, utilitarian, Rawlsian, or the ethics of rights, have revealed how and why humanity's relationship with the environment constitutes a moral problem, a problem that implies a redefinition or extension of the concepts of duty and responsibility, and an alteration in the very image human-ity has of itself and its relationship with nature. The ecological problem is first of all a problem of public *ethos*, hard to solve without questioning the ways we live together and the values held in civil society.

Economic theory is still inadequate to deal with the environmental ques-tion because of its claim to be able to solve every conflict by recourse to laws and institutions that are "neutral," i.e. that do not presuppose any adherence to values or cultural assumptions, and are thus acceptable to all actors inde-pendently of the historical context in which they are operating. The formalism of economics is also seen in the idea that a society can find its cohesion and identity in efficient "rules of the game," concerning income distribution and collective choices.

We must overcome the false antinomies between independence and belonging, efficiency and justice, self-interest and solidarity—the suppositions that the sense of belonging reduces the subject's independence, that progress in efficiency is a threat to justice, and that improvements in the individual's interest enfeeble solidarity. While need, equality, efficiency, and entitlement might plausibly have been described as competing criteria during the Indus-trial Age, these have become necessary conditions for each other in the post-industrial era. In the new regime in which human capital has become the source of value and wealth creation, need satisfaction, distributive justice, efficiency, and entitlement turn out to be complementary elements of a comprehensive approach to sustainability.

It is precisely the subject of sustainable development that today is forcing economists to rediscover the centrality of values in their scientific work. In what follows, I shall examine the emerging new interconnections between economics, ecology, and ethics.

DOI: 10.4324/9781003456940-18

Economics "Discovers" the Environment

Right from its beginnings as an independent scientific discipline, economics has focused on two central questions: how the social product is formed, and how it is distributed. The post-industrial phase of economic development has brought new problems, notably the ecological limitations that weigh on the process of production. The scarcity of resources was of course always a factor influencing the forms and rhythms of development, but up to recently scarcity was overcome by technological innovations, giving us the sense as we look back of a dizzying growth towards unlimited plenty, as if nature were no longer hostile and niggardly, as the ancients thought.

The contemporary picture is completely different. Industrial growth involves "external" effects on the environment, which were held to be negligible at the beginning of the process but are now seen to be devastating. Depletion of natural resources such as air and water may now threaten the equilibrium of the biosphere itself, perhaps definitively altered by irreversible processes. Starting from the second half of the 20th century, humanity's capacity for destruction has become a "biocidal" phenomenon. It is becoming clear that an ever-increasing production of goods and services is incompatible (*given* the known productive techniques, the present organization of the economy, and the rate of increase of the population) with safeguarding the natural and urban environment. When humanity modifies the environment too rapidly (for example transforming the seas of oil from the earth's crust into gas in the atmosphere) it creates a situation in which the speed of these changes is superior to the speed of its adaptation to them.

When public opinion began to become aware of the environmental question at the beginning of the sixties—the influence of Rachel Carson's *Silent Spring*, published in 1962, will certainly be remembered—the economists felt they were able to face up to the problem by using their own specific ways of thinking. However, the more influential shapers of public opinion passed on the idea that economics was synonymous with pollution and the destruction of nature. Economics and ecology were thus seen as opposite poles.

A major reason for this misunderstanding is that when economists in the 1960s became involved in ecological problems they thought they could make use of the instruments of analysis specifically designed for the branch of the discipline known as public economics, in its turn born of the merging of the older welfare economics and the newer theory of social choice. They saw the environmental issue as turning on resources (land, air, water, species of animals, forests) that have some basic characteristics in common: (a) they can naturally be regenerated; (b) they are often common property; (c) their over use can lead to irreversible damage, in the sense of their total exhaustion; (d) the existing stocks of these resources, and not only their flows, directly influence people's wellbeing; (e) the impact of economic activities on these resources is often cumulative and can be seen only after a certain stretch of time; (f) the environmental consequences of economic activities are basically uncertain ("hard" uncertainty in the sense that environmental uncertainty cannot be dealt with by using the tools of the familiar theory of probability).

Problems involving these resources could be addressed, the economists thought, by starting from the two central notions of public economics: externality and public good. Damage to the environment caused by economic activities could then be imputed to a typical "market failure," i.e. to the fact that in the presence of environmental resources market mechanisms no longer guarantee, on their own, the allocative efficiency that since Adam Smith was considered their most important virtue. Solutions were sought in a suitable system of taxes and subsidies, as already suggested by C. Pigou, the inventor of welfare economics.

Until recent times, economic theory has developed two main lines of research to deal with the environmental question. The first aims at devising allocative mechanisms which are both not manipulable and efficient. In this line, environmental goods are treated as factors of production. The advantage of this approach is that an externality, e.g. pollution, is merely an unaccounted-for consumption of a scarce good. This means that those inflicting an externality on others are consuming society's resources without redistributing the rent connected therewith. As long as the good is scarce, hence depletable, its consumption should be accounted for. The fact that it is not accounted for implies a sub-optional situation.

Is the Pigouvian proposal a satisfactory remedy to the problem of international externalities? Not at all, since Pigouvian taxes have never appealed to politicians or the general public. A careful examination of the emission charge and marketable permit schemes reveals that they are rarely, if ever, introduced in their textbook form. Virtually all environmental regulatory systems, using charges and marketable permits, rely on the existing permitting system. They are not implemented from scratch; rather they are grafted onto regulatory systems in which permits and standards play a dominant role.

The consequence of these hybrid approaches is that the level of cost savings resulting from implementing charges and marketable permits is generally far below their theoretical potential. Polluters have not been induced to search for a lower cost mix of meeting environmental objectives as a result of implementing charge schemes. The experience on marketable permits is similar. In other words, in order to function the economists' proposals presuppose both that a competitive set-up actually exists and that it is possible to easily monitor and enforce a system of permits and taxes. Since this is not the case, firms will prefer emission standards to emission taxes because standards result in higher profits. Emission standards serve as a barrier to entry to new firms, thus raising firms' profits. Charges, on the other hand, do not preclude entry by new firms, and also represent an additional cost to firms (see Hintermann 2013).

The second line of research is concerned with the design of political institutions that are both feasible and efficient. An institution saves on the costs of economic transactions. Therefore, rational agents, in the sense of *homo oeconomicus* rationality, will devise mechanisms in order to overcome the pitfalls of the prisoners' dilemma. Without some regulatory entity, the only alternative would be rent dissipation, leaving temporary gains to the quickest and most inefficient users. If one further assumes that the set-up cost of this entity does not use up all the captured rent—i.e. it is assumed that the "internal" transaction costs of the agency are lower than those of all single agents bargaining among each other—*and* if there is some room for repayments in the form of non-distorting lump sums, then one can conclude that the existence of an authority raises welfare in the presence of environmental goods.

Well, it is not easy to escape the feeling that we are faced here with a sort of "tin-opener" argument: suppose we have the best solution to the problem, then the problem will be solved! The truth of the matter is that it is not enough to have discovered the Pareto-improving character of the institution to be certain that it will come into existence automatically. Ascertaining the conditions for bringing such an institution into existence is the key question.

The conceptualization of the environmental problem in terms of a problem of externalities harbors a serious theoretical flaw. The notion of externality, as the effect of the action of an economic agent on the welfare of other individuals that is not captured by the price system, is not a primitive notion. It depends on the definition of economic actor and on the existence of markets. For example, if two companies operate in such a way that the one damages the other—the foundry that through its emissions of smoke damages the company nearby—an eventual merging of the two will mean that what beforehand were external effects now become a question raised within the same decision-making unit: the externality is internalized, but the pollution is still there!

It follows that we can speak of externalities only after an explanation has been provided for the number of economic actors and markets in existence. And since the number of firms and markets depends on very precise economic factors (non-convexity of production sets; transaction costs; access to information, etc.), it turns out that only an analysis of general equilibrium that, starting from market fundamentals, determined endogenously the number of firms and markets, could be a conceptually satisfying way of dealing with the question of externalities. Which it isn't, given that the two conditions that allow us to identify the existence of externality are put forward axiomatically. To give an extreme example, if only one firm existed in the economy, there could be no externality. And yet, if this firm polluted and destroyed the non-renewable resources the integrity of the environment would turn out to be damaged just the same. Among other things, this simple consideration allows us to understand why in the countries of the ex-Soviet bloc, where there was certainly no market economy, the destruction of the environment was not at all inferior to that of Western countries.

The conclusion has to be that economic science must, at the level of its very foundations, rethink the relationship between humanity and nature, leaving behind the idea of a "humanity without constraints" that leads us to believe that any devastation is legitimate, in homage to certain anthropomorphic myths of omnipotence. What is needed is the recovery of the basic recognition that humanity is part of nature, is internal to it, and has a cognitive exchange with nature, which is its necessary term. Born into it, humanity, as part of nature, changes it, which is inevitable and also useful. But this must not spell destruction.

Neither extreme anthropocentrism—which visualizes the human being as a predator—nor ecological pantheism—according to which the human species is an element of disturbance for the environment—are the solutions to the present crisis. The best foundation for environmental responsibility is the concept that the human being is the only moral subject who has responsibility for humankind, nature, and future generations.

Intergenerational Fairness and Sustainable Development

The expression "sustainable development" was originally chosen for reasons of political rhetoric. It received what we may call its official formulation in the Brundtland Report of 1987: "We mean by sustainable development a development capable of satisfying the needs of the present without compromising the capacity of the future generations to satisfy their own." But already in 1993, the Nobel Prize winner Robert Solow claimed that sustainability is a generic moral obligation of the present generation to future ones: "Insofar as it is a moral obligation, sustainability is a generic obligation, not a specific one. It is not an obligation to preserve this or that. It is rather the obligation to preserve the welfare capacity of those who come after us" (187). From this it can be deduced that the destruction of natural resources is acceptable insofar as it is compensated for by investments capable of generating other goods or services able to increase welfare. In fact, this position of Solow goes back to 1974, the year in which the American economist, inserting a non-renewable resource into a standard model of intertemporal growth, fixed a result that would afterwards become a basic reference point for the entire literature on sustainable development: a level of sustainable consumption can be guaranteed, in principle, every time it turns out to be technologically possible to guarantee a sufficient degree of substitutability between natural resource and physical capital.

For other writers, however, sustainability has to do with the property rights of future generations, an idea rendered by the phrase: "We have not inherited the earth from our parents; we are borrowing it from our children." This emotional phrase is often attributed to Ralph Waldo Emerson, though its actual origin is obscure (see Keyes 1992). In any case, this point of view is firmly shared by Howarth (1992) and Norgaard (1992) who, though accepting Solow's idea of sustainability as a question of equity between generations, do not accept its reduction to a problem of substitutability between natural resources and produced goods such as capital goods. They start here from a consideration it is easy to share, that the fact that two goods are perfect substitutes for the present generation does not imply that they are so for future generations also.

Again for other scholars, sustainability would not involve considerations about issues of distribution between generations, but considerably more traditionally, questions of economic efficiency. Starting from the premise that most environmental goods admit two alternative uses—one destructive, according to which the environment is converted into a private good enjoyed by the present generation;

and one as a public good, to be used also by future generations—Silvestre (1997) develops a model in which sustainability may be defined only in terms of the allocation of resources between generations. The interesting conclusion of the model is that, if the future generations are considered as being part and parcel of present-day society, allocative efficiency requires that environmental resources be maintained in their state of nature for a rather high number of decades.

From the end of the eighties, the awareness spread that environmental problems are global in scale, pervasive in their effects, and above all generators of important consequences for future generations. The global climatic changes, the reduction of the ozone in the atmosphere, and the irreversible damages to bio-diversity, present features that make the even quite elaborate approaches to sustainability up until that moment useless. The actions of today determine potential costs for the future generations that are inherently unforeseeable, given the dynamics and complexity of ecological systems. For example, climatic change can jeopardize the subsistence agriculture in many areas of the world, just as it may increase the frequency and the dangers of tropical storms.

It became obvious that the theoretical apparatus environmental economics had set out with was inadequate to deal with the "new" questions. Models such as Solow's, as well as the literature on the so-called optimal growth, do not face up to the question of the institutional mechanisms necessary to realize a sustainable future. In addition, it is by now obvious that social and environmental problems are closely interlinked. To be solved satisfactorily they must be dealt with together; so the assumption of *ceteris paribus* that characterizes the whole of the analysis of partial equilibrium turns out to be of very dubious usefulness (Norgaard 1993).

Some economists continue to believe that sustainability can be talked about adequately while remaining within the apparatus of cost–benefit analysis. For them, the institutions needed to ensure the internalization of environmental externalities, the efficient management of common property resources, and the efficient intertemporal allocation of resources are also sufficient to guarantee the rights of future generations. But a moment of reflection is sufficient to convince us that this is not the proper way to go about thinking of these things.

Cost–benefit analysis is very useful when we need to identify potential Paretian improvements—opportunities to improve the welfare of all without worsening the welfare of anyone. But the prices and shadow prices on which the analysis in question is based depend on the initial endowments possessed by each agent. If these are assigned in a markedly distorted way, efficiency by no means guarantees the sustainability of the development. The objective of sustainability, in other words, requires a good deal more than improvements in efficiency in the Paretian sense. It requires the carrying out of policies that enable the transfer of goods and resources from one generation to another (see Dasgupta 2008). Caring for future generations is not an altruistic concern only: improving the position of future generations enhances the future of the present generations too.

Two important consequences derive from this. In the first place, what makes the sustainability objective difficult are not just the familiar market failures, but also and above all the various forms of distributive unfairness. Secondly, the way out cannot derive from cost–benefit analysis, precisely because it possesses the tools for solving problems of efficiency but not of fairness. So the pursuit of an objective like sustainable development also means taking into consideration political and ethical aspects. To put it another way, the horizon of efficiency is not wide enough to contain the issues raised by sustainability, which is first of all a problem of the definition of the rights of different generations.

The vast literature on the subject under discussion, aside from the differences between individual writers, is founded on a shared theoretical scheme that runs as follows. On the one hand, it is assumed that all individuals are selfish, having self-interested preferences; on the other hand, that questions of fairness between generations are the concern of institutions or collective agents whose task is basically to operate transfers of resources from the present to the future generations. However, a framework of this type contains a paradox: since the social choice function on whose basis decisions at a collective level are taken is rooted in individual preferences, why should the public decision-maker, let us say a government, take responsibility for the welfare of future generations if the individuals (of which that government is the expression, and to which it answers electorally) care nothing at all about this? On the other hand, if the economic actors had solidaristic preferences towards the generations to come, what need would there be for the intervention of a government to carry out transfers of resources to the future generations?

In economics the traditional way to dissolve paradoxes of this kind is to assume that the members of present and future generations are linked to each other by bonds of a family kind that guarantee the actual transfer of goods from "parents" to their immediate descendants, i.e. their "children" (Barro 1974). This is so whenever the welfare of the children enters positively into the utility function of the parents. A way out of this kind, however ingenious, is not a great help when it comes to the problem of sustainable development, since in the temporal perspective needed to deal with the issue it is not very useful to restrict ourselves to considering only two consecutive generations. It is inconceivable that the families of the present-day generation can organize among themselves an adequate transfer of resources for the welfare of their children, who will in the future set up families in their turn. The more important transfers between the generations have to be carried through before the children have reached the stage of personal independence. It will thus be evident that it is on society as a whole that the burden falls of ensuring to future generations what is necessary to satisfy their needs.

The argument sketched above exposes a serious shortcoming in economic theory, which while it busies itself *ad abundantiam* with individual behavior and its consequences, shows no interest at all in the beliefs and motivations that lie behind human action. This gap is sometimes concealed by the consideration that, since in a market economy the consumer is sovereign and hence free to express any kind of preference, including altruistic ones, there is no reason to worry about the motivations behind his or her choices. That things do not stand like this is shown by the realization that caring for the needs of others is not an innate virtue in the human being. It is rather the result of a slow and systematic process of education. This is why for a sustainable development the argument on lifestyles that respect the creation is so centrally important.

As long as a culture founded on the models of a consumerist society prevails, especially among the young, it is obvious that politics will not be able to do otherwise than respond to this kind of signal and translate it into choices that are a logical consequence of it: increasing the levels of productivity to diminish the prices of goods and services to further increase their production and consumption, etc. C.F. Weizsäcker's words to the Seoul ecumenical assembly of 1990 are relevant here: "I know some politicians who want to do the really necessary things, but who know that as soon as they do something reasonable they will lose the next elections. It is for this reason that I am against the idea that politicians are mainly responsible, the most guilty of all. No, it is we [citizens] who are the guilty ones" (quoted in the WCC magazine *One World* 155 [May 1990]: 16).

The turbulent history of theoretical positions on environmental issues is characterized by the systematic alternation of quite markedly different points of view and lines of action. It is a history of steps forward and steps backwards, of often apparently unmotivated swings from radical innovation to conservative retreat, as if the terms of what was at stake were not clear to everyone. The fact is that without a holistic vision of the environmental issue, capable of making us realize that the environment is not simply a question of degradation or of exhaustion of resources, and without overcoming the limitations of a scientific research that is too sector-oriented and too little transdisciplinary, it will never be possible to achieve a new alliance between humankind and nature.

The Struggle Against Poverty and Sustainable Development

Where do we begin if we wish to get beyond what is still the most common, i.e. dichotomous, way of facing the crucial central problem of sustainable development? I would say that the reduction of the welfare gap between the North and South of the world is what is most crucial. The three main causes of environmental degradation are (a) the inefficient allocation of resources, (b) the iniquitous distribution of wealth and income, and (c) the disproportion between the population and the capacity of the environment to sustain it. Whereas in rich countries the first of these causes is operative, poor countries are mainly afflicted by the other two causes. Through their structural characteristics, these countries tend to specialize in the production and export of goods with a high intensity of environmental degradation. Even now, two thirds of Latin America's exports are made up of natural resources—Africa's percentage is still higher—resources that are imported and consumed in the countries of the North. These data, though crude, are already sufficient to show why the question of sustainable development cannot be separated from the reform of the rules of international trade.

When we discover that the South exports goods of a high intensity of environmental degradation, though it is not true that the South disposes of higher quantities of these goods compared to the North, we may realize why commercial policies based on the Ricardian principle of comparative advantage are a serious threat to sustainability. If we then consider that most developing countries are located in the region known as the "vital zone," characterized by highly unstable ecological equilibria and by a marked capacity to influence the atmosphere, we realize why if we continue to force these countries to use their *natural* capital to substitute for an insufficient *physical* and *human* capital, environmental degradation will inevitably accelerate.

There is still more to it than this. In 1992 the World Bank detailed the relationship existing between some indicators of environmental quality and levels of GNP per head. A relation emerged that could be shown through a curve in the form of a U turned upside down: environmental degradation grows (gets worse) with the increase of average income when the latter is at low levels, whereas it decreases (gets better) with the increase of average income when the latter has gone above a certain threshold. Basing their work on this rich empirical material, Grossman and Krueger (1994), through econometrics, found that the level of the critical threshold of average income, beyond which the above mentioned curve begins to decrease, stands at around $8,000 per head income a year (dollars of 1985). The curve in question is known in the literature as the "Environmental Kuznets curve" (EKC), from the name of the Nobel Prize winner for economics who first studied its characteristics (with reference however to the relation between levels of GNP per capita and variations of an indicator of the inequality of income within a specific population). The empirical evidence in support of the EKC is still today insufficiently robust to recommend its use for the purposes of environmental policies. It is nevertheless possible to extract from the EKC the following broad indications: some indicators of environmental degradation (emissions of CO_2; solid urban waste) increase with the increase of per capita income; others (the lack of clean water; hygiene indicators) diminish with the increase in per capita income; still others (emissions of anhydride of sulfur and nitrates) first increase and then diminish with the increase in per capita income.

What lessons can be learned from the EKC literature? Since Northern countries are to the right of the value of the critical threshold mentioned above, whereas most Southern countries are still a long way off this goal, and since the environmental problems that worry us the most today are the global ones, it is evident that we shall have to intervene urgently on the rules of international economic activities. In particular, we must realize that in the context of an increasingly globalized economy, environmental regulation and commercial regulation have to be integrated and harmonized, exactly the opposite to what has happened up until now in the WTO (see Pearson 2000).

International trade tends to separate production from consumption: an increase in the demand for tropical wood in the North translates into a corresponding reduction in tropical forest in Amazonia. Thus international trade throws a long, dark shadow over the environment. Without adequate rules and without forms of close cooperation between the agencies concerned with trade and the environment, the growing volume of commercial exchanges (in itself positive and a hopeful sign for the future) will translate into increases in environmental degradation.

The second and more important message is that the problem of the sustainability of development, under today's conditions of globalization, is intrinsically linked to the problem of poverty, both absolute and relative. It would be naive to imagine we can solve the former problem separately from the latter, or worse still, in opposition to it. Efforts to improve or conserve the quality of the environment in the North will be of very little use unless at the same time there is an urgent and comprehensive program of action against poverty to allow the countries of the South to get beyond the critical threshold identified by the ECK. Clearly, there will have to be a program of redistribution on a global scale, since policies on a national scale are no longer adequate for the purpose. If we stop and think for a moment, we find ourselves faced with a specific, yet remarkable case, in which the defense of justice also serves to improve efficiency (here identified with the sustainability of development).

It is certainly true that globalization is a positive sum game that increases aggregate wealth. But it is also true that it exacerbates the contrast between winners and losers. This fact is linked to the emergence of a new form of competition, unknown until recently: positional competition, according to which the "winner takes all and the loser loses everything"—the so-called "superstar effect" in the sense of Shermin Rose. Why is it that literature on the subject is so hotly divided? A credible answer comes from the recent work by Milanovic (2011) who distinguishes between *world* and *international* inequality. The latter considers the differences in the average incomes of various countries, unweighted ("Concept 1 inequality" in Milanovic's sense) and duly weighted to account for the size of the population ("Concept 2 inequality"). The former, on the contrary, takes into account also the inequalities in income distribution within the individual countries ("Concept 3 inequality"). It is world or global inequality which is increasing as a consequence of globalization.

In fact, in order for Concept 3 inequality to diminish, two conditions should be met: (a) poor and densely populated countries must grow at a faster rate than rich countries; (b) this must occur without an increase in inequality within the country. Now, while the first condition is more or less satisfied, the second condition is virtually absent. In fact, over the last quarter of a century, the growth rate of the poorest countries has been higher than that of the richest countries (4 percent versus 1.7 percent). Why should one worry about the growth of global inequality? One reason is because it is a principal cause of conflict and ultimately of civil war. As wisely indicated by Polachek and Seiglie (2006), conflict can be defined as "trade gone awry": if a country's gains from trade are not as high as it thinks it should receive, this becomes a major determinant of conflict. That is why the search for a socially responsible trade integration regime is a duty that the economist cannot forget about.

A related, but different, aspect is the one concerning the relationship between globalization and poverty. In the last couple of decades, poor countries have increased their participation in world trade, so much so that today they can be said to be more globalized than rich countries. Yet there is very little evidence on that relationship and even the scanty evidence available only deals with the indirect link between globalization and poverty. A notable exception is the recent work by Harrison (2006) who provides a novel perspective on how globalization directly affects poverty in developing countries. Three general propositions deserve special attention: (a) contrary to the Heckscher-Ohlin theory of international trade, the poor in countries with a lot of unskilled labor do not typically gain from trade expansion; (b) globalization generates both winners and losers among the poor and this creates social instability insofar as it destroys social capital; (c) the poor segments of population obtain the largest benefits from globalization when national governments endeavor to implement welfare policies aimed at improving the *capabilities* of life of their citizens, rather than their *conditions* of life.

Discussing the consequences of the discovery of America and the passage of the Cape of Good Hope, which he calls "the two greatest and most important events recorded in the history of mankind," Adam Smith offers a fascinating anticipation of today's arguments for a more balanced assessment of the gains and losses from cross-border exchange:

What benefits or what misfortunes to mankind may hereafter result from those great events, no human wisdom can foresee. By uniting, in some measure, the most distant parts of the world... their general tendency would seem to be beneficial. To the native, however, both of the East and West Indies, all the commercial benefits which can have resulted from those events have been sunk and lost in the dreadful misfortunes which they have occasioned... At the particular time when these discoveries were made, the superiority of force happened to be so great in the side of the Europeans, that they were enabled to commit with impunity every sort of injustice in those remote countries. Hereafter, perhaps, the natives of those countries may grow stronger, or those of Europe may grow weaker and the inhabitants of all the different quarters of the world may arrive at that equality of courage and force which... can alone overawe the injustice of independent nations into some sort of respect for the rights of one another. But nothing seems more likely to establish this equality of force than the mutual communication of knowledge and of all sorts of improvements which an extensive commerce from all countries to all countries naturally, or rather necessarily, carries along with it. (1950 [1776]: 141)

De-Growth or Development?

"Green growth" is a new term that has become the focus of much interest among policymakers concerned with enhancing both nearer-term economic progress and longer-term environmental sustainability. However, green growth differs from sustainable development in a subtle but important respect. In particular, it is not always true that green growth is good for the poor, and the poor should not be asked to pay the price for sustaining growth while greening the planet (Dercon 2014).

Would the "happy de-growth" thesis, advanced in recent times by Serge Latouche, be the proper paradigm to tackle the sustainability question? The proposal has an illustrious precedent: the theory of the stationary state initially developed by John Stuart Mill. Mill used the expression "stationary" state to project a situation where the *net* growth rate of the economy is equal to zero. Other economists and thinkers propounded analogous hypotheses in his wake. Among them I would venture to recall Nicholas Georgescu Roegen and his program of "bioecomony" in the 1970s. Therefore, we shouldn't be surprised if concerns over sustainability and the future of the planet every so often compel scholars of diverse cultural backgrounds to advance proposals like that of happy de-growth.

The social doctrine of the Church differs from this de-growth hypothesis not so much in terms of diagnosis, but rather as regards the therapy: instead of trying to treat the root causes of the illness, de-growth thinking gives in to the patient's more or less slow euthanasia. Note that the concept of development, etymologically, means "liberation from constraints" that curtail the freedom of the individual and the social aggregations in which he/she expresses him/herself. This notion of development was formulated in full at the time of civil humanism in the 15th century, with a decisive contribution from Franciscan thought. Three dimensions of human development correspond to three dimensions of liberty: the quantitative-material dimension, corresponding to freedom *from*; the social-relational dimension, corresponding to freedom *to*; and the spiritual dimension, corresponding to freedom *for*.

It is obviously true that as conditions stand today the quantitative material dimension overrides the other two, but this by no means bestows legitimacy on the conclusion that reducing (or nullifying) growth—which regards the material dimension—would foster progress on the part of the other two dimensions. In fact, it can be demonstrated that exactly the contrary is true. This is why the social doctrine of the Church (and especially Benedict XVI's *Caritas in veritate*) speaks about integral human development, about development which must maintain harmony and mutual equilibrium among the three dimensions. This takes place through a change in the *composition*—and not the *level*—of the basket of consumer goods: fewer material goods, more relational goods, more immaterial goods.

Therefore, the antidote to the current consumeristic model is not de-growth, but rather the civil economy, a typically Italian program of research and thought dominant throughout Europe until the end of the 18th century, and since then overclouded by the program of political economy. Take note of the differences: while the civil economy pursues the common good, the political economy pursues the total good. While the latter considers it possible to resolve problems in the economic-social realm on the basis of the principles of the exchange of equivalents and redistribution driven by the state, the civil economy flanks these two principles with the principle of reciprocity, which is the practical precipitate of fraternity. The original contribution of *Caritas in veritate* (chapter 3) is that it restored to fraternity that central role in the economy which had been completely wiped out by the French Revolution and Bentham's utilitarianism.

Humanize the market, don't demonize it: this is the slogan that describes the challenge confronting us today. As paradoxical as this may appear, the thesis of degrowth does nothing more than add a minus sign to the standard paradigm of political economy, and in no way constitutes progress beyond or above it. This is why it cannot be considered a solution for the many and grave problems now afflicting our respective societies. The social doctrine of the Church will never be able to accept any regression: those who cultivate the concept of time as *kairos*, and not merely as *chronos*, know that difficulties are surmounted by changing one's outlook towards reality—and not with operations that would wind the clock of history backward.

The Crucial Role of Virtues

Given the problems and the difficulty of solving them, should we just let the processes occurring today go ahead according to their own internal logic? That would be overwhelmingly irresponsible, because in actual fact there is no need, as some people suggest, to halt the process of growth or that of globalization. What is really needed, and urgently, is to work for the establishment of an economic and social order founded on the plurality of power centers, i.e. on a polyarchy, which unlike pluralism, is not just a question of numerousness, but of diversity—both of the modes of production and patterns of consumption.

Above all, we need to recover the sense of responsibility. It is true that the concept of responsibility, today, meets many difficulties in being accepted, let alone applied. On one hand, globalization is increasing, in unprecedented ways, the distance between action and the ultimate consequences of the action. Consider the impact of processes of mergers and acquisitions on the phenomenon of "short-termism": firms fearing take-overs tend to pay scant attention to whatever does not have a return in the short-run—including social responsibility. On the other hand, the new technologies that connote the third industrial revolution tend to reduce the sense of responsibility insofar as they increase the number and typology of the unpredictable consequences of actions. The notion of responsibility is strictly connected to that of accountability. Responsible is s/he who knows how to manage situations, adequately evaluating their risks and results. But the current technological changes render this exercise ever more difficult, if not impossible.

That is why we find ourselves obliged to turn to ethics. But which ethical theory is adequate to the purpose? My answer is the ethic of virtues, which Adam Smith—developing a line of thought going back to the civil humanists of the 15th century, elaborated in *The Theory of Moral Sentiments* (1759). The institutional structure of society—says Smith—must favor the dissemination among citizens of the civic virtues. If economic agents don't already embody in their structure of preferences those values that they are supposed to respect, there isn't much to be done. For the ethic of virtues, in fact, the enforceability of the norms depends, in the first place, on the moral constitution of individuals. It is because there are stakeholders that have ethical preferences—that attribute, that is, value to the fact that the firm practices equity and works for the dignity of people *independently* of the material advantage that can be derived—that the ethical code can be respected *also* in the absence of the mechanism of reputation.

The point worth highlighting in particular is that the key to the ethic of virtues is in its capacity to resolve the opposition between self-interest and interest for others, between egoism and altruism, by moving beyond it. It is this opposition, child of the individualistic tradition of thought, that prevents us from grasping that which constitutes our own wellbeing. The virtuous life is the best not only for others—as the various economic theories of altruism would have it—but also for us. This is the real significance of the notion of common good, which can never be reduced to a mere sum total of individual wellbeings. Instead, the common good is the good of being in common. That is, the good of being inserted into a structure of common action, which is exactly what is required in order to sustain nature.

Common is the action that, in order to be carried out, requires both the *intentional* coming together of many subjects (and of which all the participants are aware) and of intersubjective relationships that lead to a certain unification of efforts. More precisely, there are three elements that distinguish a common action. The first is that it cannot be concluded without all those who take part being conscious of what they are doing. The mere coming together or meeting of many individuals is not enough. The second element is that each participant in the common action must retain title, and therefore responsibility, for that which he does. It is exactly this element that differentiates common action from collective action. In the latter, in fact, the individual's identity disappears and with him disappears also personal responsibility for that which he does. The third element is the unification of the efforts on the part of the participants in the common action for the achievement of the same objective. The interaction among many subjects in a given context is not yet common activity if they follow diverse or conflicting objectives.

Now we can appreciate the specific value that the ethic of virtues offers, that of liberating us from the obsessive Platonic idea of good, an idea that says there is an *a priori* good from which an ethic is extracted to be used as a guide to our actions. Aristotle—the initiator of the ethic of virtues—in total disagreement with Plato, indicates for us instead that the good is something that happens, that is realized through activities. As Lutz (2003) puts it, the most serious problem with the various ethical theories stemming from the individualistic tradition of thought is that they are not capable of offering a reason for "being ethical." If it's not good for us to behave ethically, why do what is recommended by ethics? On the other hand, if it is good for us to "be ethical," then why would it be necessary to offer managers incentives for doing that which is in their own interest to do? The solution to the problem of moral motivation of decision makers is not that of setting constraints (or providing incentives) for acting against their self-interest, but to offer them a more complete understanding of their own wellbeing. Only when ethics becomes part of the objective function of the agents does moral motivation cease to be a problem, because we are authentically motivated to do that which we believe is best for ourselves. This is why cultivating civic virtues is the undeniable task not only from the point of view of citizenship—something known for a long time—but also from the point of view of sustaining nature.

Economics is inextricably a part of ethics because humans are not aloof islands of exchange; rather, they live, work and thrive in social settings. Humans have innate dispositions *for* self, *for* others, and *against* others that serve useful functions, yet whose claims must be internally adjudicated by a moral agent. Understanding individual and social conceptions of "right" and "wrong" is essential for the environmental problematic. When one acknowledges the looming crisis of our civilization one is practically obliged to abandon any dystopic attitude and dare to seek out new paths of thought. As T.S. Eliot once observed, you can't build a tree; you can only plant one, tend it, and wait for it to sprout in due time. You can, however, speed up its development with proper watering. For, unlike animals, which live in time but have no time, human beings have the ability to alter their times.

Note: Previously published in *The Japan Mission Journal*, 2012.

References

Barro, R. (1974). "Are government bonds net wealth?" *Journal of Political Economy* 82: 1095–1117.

Brundtland, G.H. (1987). *Our Common Future: Report of the World Commission on Environment and Development*. Geneva: UN-Dokument A/42/427.

Dasgupta, P. (2008). "Discounting climate change." *Journal of Risk and Uncertainty* 37: 141–169.

Dercon, S. (2014). "Is green growth good for the poor?" *The World Bank Research Observer* 29: 163–185.

Grossman, G.M., and A.B. Krueger (1994). "Economic growth and the environment." NBER, WP 4634.

Harrison, A. (2006). "Globalization and poverty." NBER, WP 12347, June.

Hintermann, B. (2013). "Market power in emission permit makets: Theory and evidence!" CESifo, 4447, Oct.

Howarth, R. (1992). "Environmental valuation under sustainable development." *American Economic Review* 82: 473–477.

Keyes, R. (1992). *Nice Guys Finish Seventh: False Phrases, Spurious Sayings, and Familiar Misquotations*. New York: Harper Collins.

Lutz, D. (2003). "Beyond business ethics." *Oikonomia* 2: 13–28.

Milanovic, B. (2011). *The Haves and the Have-Nots*. New York: Basic Books.

Norgaard, R. (1992). "Sustainability and the economics of assuring assets for future generations." *WPS* 832. Asia Regional Office: The World Bank.

Norgaard, R. (1993). "The co-evolution of economic and environmental systems and the emergence of unsustainability." In R.W. England, ed. *Evolutionary Concepts in Contemporary Economics*. Ann Arbor, MI: University of Michigan Press.

Pearson, C.S. (2000). *Economics and the Global Environment*. Cambridge: CUP.

Polachek, S. and Seiglie, C. (2006). "Trade, peace and democracy: An analysis of dyadic dispute," IZA, DP 2170, June.

Silvestre, J. (1997). An Efficiency Argument for Sustainable Use. In J.E. Roemer, eds, *Property Relations, Incentives and Welfare*. London: Palgrave Macmillan. https://doi.org/10.1007/978-1-349-25287-9_2

Smith, A. (1950 [1776]). *The Wealth of Nations*, ed. E. Cannan. London: Methuen.

Solow, R. (1993). "Sustainability: An economist's perspective." In R. Dorfman and N. Dorfman, ed. *Economics of the Environment*. New York: Norton.

World Bank (1992). *World Development Report 1992: Development and the Environment*. Washington, DC: The World Bank.

Chapter 14

Ecology and Economy

Rowan Williams

The True Measure of Wealth

The two big "e-words" (ecology and economy) in my title have sometimes been used in recent decades as if they represented opposing concerns. Yes, we should be glad to do more about the environment, if only this didn't interfere with economic development and the liberty of people and nations to create wealth in whatever ways they can. Or perhaps, we should be glad to address environmental issues if we could be sure that we had first resolved the challenge of economic injustice within and between societies. So from both left and right there has often been a persistent sense that it isn't proper or possible to tackle both together, let alone to give a different sort of priority to ecological matters.

But this separation or opposition has come to look like a massive mistake. It has been said that "the economy is a wholly-owned subsidiary of the environment." The earth itself is what ultimately controls economic activity because it is the source of the materials upon which economic activity works. Increasingly, economists have expressed unease about the habit of thinking of environmental matters as "externalities" where issues about economic development are concerned; and Professor Partha Dasgupta of Cambridge has argued very cogently that we need to stop measuring wealth in terms of GNP and to include reference to human and natural capital in any serious measure of national well-being. It is perfectly possible for a country to show an increase in its GNP and even its Human Development Index, and in fact to be experiencing overall economic decline because of the erosion of natural resources and the rate of population growth. In a paper of 2002, Dasgupta demonstrated that even in the Indian subcontinent, often cited as a good news story for gradual wealth accumulation, the pattern is really one of decline in the light of these factors. In Pakistan, for example, GNP figures suggest that the national economy grew at a steady annual rate of 2.7 percent between 1965 and 1993. But when depletion of natural assets and population growth are factored in, it appears that the average Pakistani actually became poorer during that period.

To say this is to identify the tip of an iceberg. And the bulk of that iceberg is our incapacity to develop a view of economics that takes account of a sufficient range of factors for really dependable prediction.

DOI: 10.4324/9781003456940-19

Depletion

The pattern we currently see in the world economy is a sort of pincer movement in respect of natural resources. We are taking resources out of the biosphere; and we are contributing to the biosphere a set of lethally dangerous extras. Both can be illustrated with one example. Contemporary methods of fish farming (aquaculture) require large quantities of wild fish as food for farmed fish, so that there is a further dramatic depletion in the wild fish populations. Fishermen who still depend on wild fishing have to pull in greater quantities to compete with farmed produce. Farmed fish contain higher levels of toxins than wild fish; if they escape, as they often do, they interbreed with wild fish and introduce those toxins to the wild strains—as well as introducing genetic complications, since farmed fish, bred for rapid growth, have poor survival capacity outside controlled conditions (see Diamond, 2005, p. 488). Thus we simultaneously deplete and poison. It is an elegant metaphor for a very wide range of phenomena. The impact of carbon emissions is now so well-known that it is hardly worth rehearsing the problems—though the most recent scientific summit on this matter convened by the government at Exeter some few weeks ago agreed, to the dismay of many, that practically all of our estimates of damage in terms of climate deterioration had been spectacularly overoptimistic. The measurable rise in temperature—and the hitherto underrated extent of acidization in the ocean through carbon pollution—left little doubt that the predicted rise in water levels would be substantially greater within the next decade, and that life in the oceans was more at risk than had been realized. And of course to speak of rising water levels is not just to predict a gentle advance; melting at or near the poles could mean vast slippages of ice capable of triggering a tsunami effect. We know today what that can mean in a way that we could hardly have dreamed of even a year ago. But we are not only speaking about carbon emissions; we know something of the effects of pesticides and herbicides, and we have become more acutely conscious of the chemical cocktails in our food and water. And the transfer, for economic reasons, of plant and animal species from one environment to another has had a regularly devastating effect on the overall ecology of a new environment and its balance.

Economy and ecology cannot be separated. If Dasgupta is right about the proper definition of wealth, ecological fallout from economic development is in no way an "externality"; it is a positive depletion of real wealth, the "human and natural capital" of which he speaks. We should not be surprised; after all, the two words relate to the same central concept. An *oikos* is a house, a dwelling-place: ecology is the science of what makes up a dwelling place, an environment, the way it works and holds together, the "logic" of a material setting; and economy is the law that regulates behavior in an environment, the active "housekeeping" that manages what is at hand. To seek to have economy without ecology is to try and manage an environment with no knowledge or concern about how it works in itself—to try and formulate human laws in abstraction from or ignorance of the laws of nature. Much of what I have been saying so far is indebted to the new study by the American biologist and geographer Jared Diamond; and he offers a vivid image for the nature of our ignorance. Why, people ask, should we be bothered about the survival of "lousy little species" that appear to have no use? "The entire natural world," Diamond replies, "is made up of wild species providing us for free with services that can be very expensive, and in many cases impossible, for us to provide for ourselves. Elimination of lots of lousy little species regularly causes big harmful consequences for humans, just as does randomly knocking out many of the lousy little rivets holding together an airplane" (489).

Troubling Signs

We cannot continue to pretend to "keep house" for the human race if we refuse to pay any attention to where in the house the gas pipes and electricity wires are laid, which walls are supporting walls, or where the water is carried off by the guttering. But how many sentences in lectures on this subject have begun with the words, "We cannot continue to..."? Hundreds at least; apparently we can in the short term. But the most original and disturbing aspect of Diamond's book is its remarkably wide-ranging demonstration that failure to manage the environment is a major decisive factor (though admittedly not the only one) in the collapse of settled cultures throughout the centuries, from Easter Island to Viking Greenland; and that collapse is—as with the rising of water levels—no gentle decline but a bloody and costly disintegration. Diamond applies his model not only to the distant past but also to the history of the genocide in Rwanda and Burundi. We know a little about the way in which economic "rationalization" to meet the requirements of the World Bank at the end of the eighties put pressure on Rwanda, contributing to the social rootlessness that led to militarization. We are conscious of the poisonous legacy of colonial manipulation of tribal rivalry (perhaps an issue to which Diamond gives insufficient attention, as at least one reviewer has argued). But we are only slowly recognizing the role of population growth, environmental degradation, and consequent land shortage in fueling the conflicts that followed. Such problems cannot indefinitely drift on; "sooner or later they are likely to resolve themselves, whether in the manner of Rwanda or in some other manner not of our choosing, if we don't succeed in solving them by our own actions" (328).

Social collapse is a real possibility. When we speak about environmental crisis, we are not to think only of spiraling poverty and mortality, but about brutal and uncontainable conflict. An economics that ignores environmental degradation invites social degradation—in plain terms, violence. It is no news that access to water is likely to be a major cause of serious conflict in the century just beginning. But this is only one aspect of a steadily darkening situation. Needless to say, it will be the poorest countries that suffer first and most dramatically, but the "developed" world will not be able to escape: the failure to manage the resources we have has the same consequences wherever we are. In the interim—just as within so much of urban society in wealthier countries—we can imagine "fortress" situations, struggling to keep the growing instability and violence elsewhere at bay and so intensifying its energy. We can imagine increasing levels of social control demanded, with all that that means for our own internal harmony or stability. And we are not talking about a remote future. There are arguments over the exact rates of global warming, certainly, and we cannot easily predict the full effects of some modifications in species balance. But we should not imagine that uncertainty in this or that particular area seriously modifies the overall picture. On any account, we are failing.

New Technologies to the Rescue?

It is relatively easy to sketch the gravity of our situation; not too difficult either to say that governments should be doing more. Government is crucial, and it matters a great deal that the UK administration has declared a commitment to action on climate change. But governments depend on electorates; electors are persons like us who need motivating. Unless there is real popular motivation, governments are much less likely to act or act effectively; there are always quite a few excuses around for not taking action, and, without a genuine popular mandate for change, we cannot be surprised or outraged if courage fails and progress is minimal. Our own responsibility is to help change that popular motivation and so to give courage to political leaders. And this means challenging and changing some of the governing assumptions about ourselves as human beings.

One of the reasons sometimes given for not being too alarmed by predictions of ecological disaster is that we are underrating the possibilities that will be offered by new technologies. Thus the American economist Nancy Stokey, responding to a very detailed discussion by another American economist, William R. Cline, of the impact of climate change and the measures necessary to control it, describes Cline's picture as "alarmist": "he makes no allowance for technical change in the next 300 years that will allow the world to cope more effectively with CO2 emissions and their climatic effects" (Lomborg, 2004, p. 642). Apart from the assumption that we have time to spare in this matter, what is startling is the appeal to "technical change" in these general terms as a messianic resource. Diamond notes at the end of his book that technical changes introduced to solve environmental problems have a spectacular record of generating fresh problems (he instances the motor car and the development of CFC gases as safe refrigerating agents; 505–506). If we simply do not know what "technical change" might lie ahead and if the history of technological "fixes" is so unpromising, it takes a great deal of blind faith to think that we can soften the projections of danger in this way. And if this is so, one of the areas in which we have to challenge assumptions is in this matter of reliance on technology to solve problems that are actually about human choices.

To appeal to a technical future is to say that our most fundamental right as humans is unrestricted consumer choice. In order to defend that, we must mobilize all our resources of skill and ingenuity, diverting resource from other areas so that we can solve problems created by our own addictive behaviors. The question is whether, even if this were clearly possible (which is anything but clear; you can't solve a challenge like this with the mere confidence that something will turn up), it would be a sane or desirable way of envisaging the human future. There would always be a case for putting the technical response to new crises ahead of other human needs—since we should always have to ensure we had an environment at all. But this sounds suspiciously like a recipe for perpetuating anxiety and even injustice; we ought not, surely, to be taking for granted that it is a future to be aimed at. It has been said more than once that a future of tighter technical control is also likely to be one of tighter human control. It is not as if we could simply contemplate a libertarian paradise.

But if this is so, there is no alternative to challenging the underlying motivation. Dasgupta would have us redefine wealth itself in a way that relativizes GNP and includes the idea of natural capital; can the same kind of redefinition apply to our ideas about individual wealth or security? What if we believed that the wealthy or secure person was one whose relationship with the environment was one in which actual enjoyment of and receptivity to the environment played the most significant part? This suggests something of a paradox. In order fully to access, enjoy, and profit from our environment, we need to see it as something that does not exist just to serve our needs. Or, to put it another way, we are best served by our environment when we stop thinking of it as there to serve us. When we can imagine what is materially around us as existing in relation to something other than our own purposes, we are free to be surprised, educated, and enlarged by it. When we obsessively seek to guarantee that the environment will always be there for us as a storehouse of raw materials, we in fact shrink our own humanity by shrinking what is there to surprise and enlarge, by reducing our capacity for contemplation of what is really other to us.

What the Religions Have to Say

All the great religious traditions, in their several ways, insist that personal wealth is not to be seen in terms of reducing the world to what the individual can control and manipulate for whatever exclusively human purposes may be most pressing. Judaism's teachings about the "jubilee" principle stress that the land is lent not given to human cultivators: it requires "sabbatical" years, and its value is to be seen not in terms of absolute possession but as a source of a limited number of harvests between the sabbatical years (Lev 25). The assumption is that the environment that is given, the land bestowed by God, has to be set free regularly from our assumption that it belongs to us; it has to be left to be itself, to be in relation simply to the God who has given it. A year of uncultivation, and wildness, is not a lot, but it speaks eloquently of our willingness to organize economy around ecology, to "keep house" within the limits of a world where we are guests more than owners. Similarly, Christianity not only has its challenges in the Sermon on the Mount to anxiety about controlling the environment, prohibiting us from identifying wealth with possession; it also has its sacramental tradition which presents the material order as raw material for the communication of God's love—the

Eucharist as the effective symbol of God's action in creating a radically different human society, not characterized by rivalry and struggle for resources. At the center of Christian practice is a rite in which all are equally fed by one gift, and in which material things are identified symbolically with the self-offering of Christ. Islam also underlines the partnership of humanity and the rest of the natural order—and, in a passing observation in the Qur'an (Sura 16.8) reminds us that some of the purposes of the animal creation are unknown to us. And a twentieth-century Iranian scholar (Muhammad Husayn Tabatabai) quotes both Muhammad and the fifth of the Shi'a Imams as commending farming because it is beneficial for humans and for the animal world as well. Examples could be multiplied from these and other faiths; but what I have quoted makes it abundantly clear that religious faith assumes that our humanity grows into maturity by allowing the material environment its own integrity. While the detail of this is inescapably complex, the point is plain. The *oikos* we inhabit has a *logos*, a meaning whose fullness is not exhausted in what we can make of it; the *nomos*, the law of our behavior in this dwelling-place, has to work with and not against the larger significance of a world that stands first in relation to its maker, and so has to be seen as free from our preoccupations about its usefulness to us.

The jubilee idea has had great currency recently as a focal image for the imperative of debt remission; I believe it has just as much importance in this context—and indeed that using it in this context reminds us of the way in which the issues of economic justice and of ecological justice belong together. Perhaps we need another "jubilee" campaign, concentrated on sabbaticals for overfished waters and deforested uplands, recognizing that the rapacity and short-term planning that devastate these resources have their roots in the same blindness that, three decades ago, began to press disadvantaged nations into debt and then sought to improve their economies by the profoundly damaging strategies of "structural adjustment," which deplete the human—the civil and cultural—resources of a nation.

The unique contribution that can be made to this whole discussion by religious conviction might be characterized in two ways. Religious belief claims, in the first place, that I am most fully myself only in relation with my creator; what I am in virtue of this relationship cannot be diminished or modified by any earthly power. It is this that grounds the obstinate belief in the irreducible value of human persons which animates any religious witness or work for the sake of justice; it is this that enables religious resistance to even the most overwhelmingly powerful and successful tyrannies, from the Roman Empire to the Third Reich, the Soviet Union or apartheid South Africa. But the implication, secondly, is that every aspect of creation likewise finds its full value and significance in relation to the creator, not to the agenda of any other creature. In the environment there is a dimension that resists and escapes us: to be aware of that is to grasp the implications of belief in human dignity, in my own dignity or value. And to reduce the world to a storehouse of materials for limited human purposes is thus to put in question any serious belief in an indestructible human value. As writers like Mary Midgley have argued eloquently, humanity needs to rejoin the rest of creation, to become aware of the limits that interdependence imposes and of the dangerous groundlessness of belief in human value when it is abstracted from a sense of value in all that exists around us.

New Structures

We are speaking about redefining wealth as "wealth that builds and sustains and takes forward the core purpose of our whole human enterprise" (Zohar and Marshall, 2004, p. 33). If we have to use the language of rights here—and it is ambiguous in many ways—we ought to be saying that human persons have a right to live in an environment that is not only safe and healthy in the obvious sense but also is itself, not fully determined by human projects. We could imagine a "charter" of rights in relation to the environment—that we should be able to live in a world that still had wilderness spaces, that still nurtured a balanced variety of species, that allowed us access to unpoisoned natural foodstuffs. Over the twentieth century, there have in fact been a good many moves in such a direction—in the UK through clean air legislation and the maintenance of public parks and the work of many conservation trusts. It may be that the time is ripe for an attempt at a comprehensive statement of this, a new UN commitment—a "Charter of Rights to Natural Capital" to which governments could sign up and by which their own practice and that of the nations in whose economies they invested could be measured. But we should make no mistake: the possibility of anything like this depends on each of us. Already consumer power has begun to make a difference to the practices of international business in pressing for signs of environmental responsibility; governments need to strengthen their commitments and need electoral incentives to be involved in the sort of internationally agreed aspirations I have sketched.

But aspirations alone are of no use. We return constantly in discussions of this subject to what sort of structures and sanctions might assist in making effective a change in our motivations and myths. A charter may be desirable, but it needs institutional backing. Various suggestions have been advanced; and it is worth noting that very different commentators have come to convergent views on the sort of thing that is required. Sir Crispin Tickell has argued in a lecture last year for a "World Environment Organization" comparable to the World Trade Organization and capable of working in harness with it. George Monbiot has elaborated the model of a "Fair Trade Organization" that would establish both ecological and economic standards for multinational trading. It would act as a global licensing body, restricting trade and enterprise across national boundaries to those companies that were ready to abide by a set of specified criteria at every stage of their activities. "If, for example, a food-processing company based in Switzerland wished to import cocoa from Côte d'Ivoire, it would need to demonstrate to the Fair Trade Organization that the plantations it bought from were not employing slaves, using banned pesticides, expanding into protected tropical forests, or failing to conform to whatever other standards the organization set" (Monbiot, 2003, p. 228). As he points out, there are already examples of such regulatory regimes in operation, some voluntary (as with the existing fair trade movement), some mandatory, such as health and safety regulations within the jurisdiction of individual nations. Is it impossible to think of internationally enforceable regulation of this sort? Monbiot goes so far as to float the possibility of expanding the remit of the International Criminal Court to deal with companies that distort or bypass the liberties of elected governments in forcing environmentally and socially disastrous developments on them (p. 230)—a drastic course of action, which would bring its own complications; but the idea itself at least underlines the sense in which environmental disaster can be as destructive as military crimes.

We are looking here at new sorts of structures. Yet through institutions like the WTO, we already see possibilities. Whether a new regulatory body should be a partner to the WTO or should be a comprehensive body dealing with the large issues Monbiot outlines, a sort of combination of WTO and a "World Environment Organization," matters less than the willingness to entertain and acknowledge the urgency of some intensified international regime to monitor and discipline economic activity in the ways we have been discussing. A manageable first step relating particularly to carbon emissions, supported by a wide coalition of concerned parties, is of course the "Contraction and Convergence" proposals initially developed by the Global Commons Institute in London. This involves granting to each nation a notional "entitlement to pollute" up to an agreed level that is credibly compatible with overall goals for managing and limiting atmospheric pollution. Those nations which exceed this level would have to pay pro rata charges on their excess emissions. The money thus raised would be put at the service of low emission nations—or could presumably be ploughed back into poor but high-emission nations—who would be, so to speak, in credit as to their entitlements, so as to assist them in ecologically sustainable development.

Such a model has the advantage that it seeks to intervene in what is presently a dangerously sterile situation. At the moment (2005), some nations that are excessive but not wildly excessive polluters (mostly in Western Europe) have agreed levels of reduction under the Kyoto protocols, and are moving with reasonable expedition towards their targets; some developed nations that are excessive polluters have simply ignored Kyoto (the USA); some rapidly developing nations that are excessive polluters have also ignored Kyoto because they can see it only as a barrier to processes of economic growth already in hand (India and China). A charging regime universally agreed would address all these situations, allowing the first category to increase investment aid in sustainable ways, obliging the second to contribute realistically to meeting the global costs of its policies, and enabling the third to explore alternatives to heavy-polluting industrial development and to consider remedial policies.

This scheme deals with only one of the enormous complex of interlocking environmental challenges; but it offers a model which may be transferable of how international regimes may be constructed and implemented. If Contraction and Convergence gained the explicit support of the UK government, this would be a significant step towards political plausibility for the program, and it is well worth keeping the proposals in the public eye with this goal in mind. Election campaigns seldom give much space to environmental matters; but the perceived significance of these concerns is weightier now than it has ever been, and the UK's declared commitments on climate change provide an important lever for bringing them into fuller focus as we move towards the election. Just as in the realm of consumer pressure, it is up to us how high a profile a plan such as Contraction and Convergence has in the questions we raise for political candidates.

Guests of the Earth

But it is because the ecological agenda is always going to be vulnerable to the pressure of other more apparently "immediate" issues that it cannot be left to electoral politics alone. Change in consumer attitudes, leading to the gradual emergence of slightly more eco-friendly policies on the part of major retailers, did not happen primarily as a result of conventional political activism, but in the wake of a persistent drip-feed of information and the identification of simple local means of exercising consumer power. As Jared Diamond says in an appendix to his book, the most effective action occurs when people have worked out the point in the commercial chain where they can most constructively bring pressure to bear: "Consumers... need to go to the trouble of learning which links in a business chain are most sensitive to public influence, and also which links are in the strongest position to influence other links" (2005, p. 557). Consumer pressure (for abundant energy sources, for fast food, for efficient refrigeration, for rapid travel and so on) has always been a major part of the problem in the development of ecologically irresponsible economics; the question is now whether it can be part of the solution.

The indications certainly are that it can. But in a context where information overload makes us rapidly bored or disoriented or both, we still need a steady background of awareness and small-scale committed action, nourished by some kind of coherent vision. Ecologists have argued regularly that some religious attitudes are part of the problem; once again we have to ask whether religion is part of the solution. Certainly, what has sometimes been said about the responsibility of the Judeo-Christian tradition for the exploitation of the earth is a caricature, in the light of the theological resources touched on earlier in this lecture; nor is it true that pre-modern or non-Western societies innately possess a superior wisdom that delivers them from ecological follies. But there is this amount of truth in the caricature: the alliance of early modern Western culture in its first flush of energy—eagerly problem-solving, expansionist, colonialist, functionally minded—with a certain kind of Christianity—triumphalistic, rational, and unsympathetic to the idea of a sacred world of symbolism, heavily focused on ideas rather than acts and relations—has undoubtedly been a factor in what is so often called the "disenchantment" of the natural environment. The slow rediscovery, in and out of the Christian fold, of that dimension of the environment that is in no way defined by its relationship with us but exists in its own relationship with God has posed a proper and grave challenge to what is left of the early modern rationalist/expansionist alliance.

But it is an open question whether either a simply secular philosophy or a diffuse "sense of the sacred" in the environment will fully do the job. In these reflections, we have come back more than once to the question of how we define wealth. The historic religious traditions see it, in one way or another, as bound up in relation with an entire environment that is understood as given "in trust"; we are answerable in respect of our relation with the material world, as we are answerable for what we make of ourselves. This is more than just an awareness of "sacred" depth in things; it is recognizing that we are bound to be involved in the intervention in our environment, but that this intervention has to be measured by something more than the meeting of our needs. Thus religious faith steers us away from any fantasies we may have of not "interfering" with the environment (the first planting of grain was an interference), but it tells us that our interaction with what lies around can never be simply functional and problem-solving. We have to discover a way of preserving an environment whose freedom from our anxious and exploiting need becomes a vital contribution to our own lives and our sense of our dignity. In honoring the freedom of what lies around us to be more than a storehouse for our gratification, we give the respect that is due to environment as creation—and thus give due honor to a creator whose purposes are not restricted to what we can grasp as good for us alone (remember the important reservation in the Qur'anic text I quoted about the unknown purposes of God in the animal creation).

Wealth is access to the "capital" of the world as it is, access to the truth and reality that can be discovered when we are set free from our narrow and self-directed concerns—a discovery that both individuals and societies need to make. As such it is access to the depth of our own being, to the rich capacity of the world around to generate in us joy and amazement as well as practical sustenance, and to the final depth of reality which is the love of God as the source of all gifts. We shall not be able adequately to deal with our crisis of "housekeeping" without what I earlier called the sense of being a guest in the *oikos* of our world, the sense that ought to keep together the logic of the household and the discipline of the household, ecology, and economy. Religious commitment becomes in this context a crucial element in that renewal of our motivation for living realistically in our material setting—the motivation that is vital if we are to avoid the collapse of civil discourse, material sustainability, justice, and stability which, if Diamond is right, regularly accompanies ecological degradation. The loss of a sustainable environment protected from unlimited exploitation is the loss of a sustainable humanity in every sense—not only the loss of spiritual depth but ultimately the loss of simple material stability as well. It is up to us as consumers and voters to do better justice to the "house" we have been invited to keep, the world where we are guests.

Note: Previously published in *The Japan Mission Journal*, 2012.[1]

Note

1 Rowan Williams, 2012, *Faith in the Public Square*, Bloomsbury Continuum, an imprint of Bloomsbury Publishing Plc.

References

Dasgupta, Partha (2002). "Is contemporary economic development sustainable?" *Ambio* 31(4): 269–271.

Diamond, Jared (2005). *Collapse: How Societies Choose to Fail or Succeed*. New York: Viking.

Lomborg, Bjørn, ed. (2004). *Global Crises, Global Solutions*. Cambridge: Cambridge University Press.

Monbiot, George (2003). *The Age of Consent: A Manifesto for a New World Order*. London: Harper Perennial.

Zohar, Danah, and Ian Marshall (2004). *Spiritual Capital: Wealth We Can Live By*. San Francisco, CA: Berrett-Koehler.

Chapter 15

Atmospheric Intervention

Lauren Guilmette

This essay began in the process of building a community garden in the US South, working with neighbors to construct a shared public space in an underserved neighborhood traumatized by gun violence. My thinking started from a recurring observation from a community leader that there was something in the atmosphere: an unresolved, haunted energy. In the essay that follows, I ask, first, what does it mean to feel something in the atmosphere? To this end, I turn to recent work on atmospheres and affective commons, and I argue that this atmospheric turn would be deepened by closer engagement with the late Australian feminist Teresa Brennan (1952–2003).[1] Second, I ask how and for whom is it right to intervene, to try to steward this atmosphere? I find "intervene" less moralizing than "fix" or "improve"; still, it maintains a sense of grandeur that gives me pause.

It first bears clarifying: what do I mean by an atmosphere? How to describe this transpersonal, more-than-human mood, at once sociohistorical and ecological? In phenomenology as well as in geography, gender studies, and other discourses, "affective atmospheres" name a shared emotion or mood existing among and between sentient beings rather than just within them.[2] Kristen Simmons writes of "settler atmospherics" oppressing indigenous communities,[3] while Eric Stanley draws this thinking about atmospheres to conditions of livability for queer and trans people.[4] Sara Ahmed writes of atmospheric walls, while Jack Leff explores the circulation of tear gas by state forces in response to protests.[5] In line with their earlier work on precarity, Judith Butler writes of the ambivalence of shared air, in which the risks and harms are not equally distributed: "The pandemic upends our usual sense of the bounded self, casting us as relational, interactive, and refuting the egological and self-interested bases of ethics itself."[6] Butler's new book brings this affective sense of atmosphere into dialogue with environmental racism and the differential dumping of waste products which, insofar as they continue to exist, unfortunately must go somewhere.[7] Similarly, Jean-Thomas Tremblay explores breathlessness and this racialized distribution of smog and toxins in BIPOC communities.[8] Lauren Berlant writes of the infrastructures undergirding and reproducing atmospheres as "generative, multiple, and often contested processes involved in the substantive connections among people and lifeworlds."[9] According to Berlant, these processes

DOI: 10.4324/9781003456940-20

reveal our "nonsovereign relationality," an inconvenient reliance upon others, from the systems that provide our basic needs to the air we breathe.

It is an unfortunate trend that Teresa Brennan's work tends to be reduced, in this emerging discourse, to the over-quoted line about "feeling the mood in a room" that opens *The Transmission of Affect* (2004).[10] Thus, Sara Ahmed groups Brennan with theorists of "affect contagion" who minimize the ways affects are misinterpreted in transmission, and she takes Brennan to suggest we all feel that mood in the same way, regardless of positionality.[11] Again taking distance from Brennan's thought but for different reasons, Berlant writes, "It's not just what Teresa Brennan describes as the affective discernment one's gut performs when entering a room or a thing that grabs your attention in a way in which you're fully present, but... sensed forced entailment, like being drafted."[12] Does Brennan attribute too much agency to our attentive capacities and thereby underestimate the force of an atmosphere? Berlant's critique sits at odds with Ahmed's concern that Brennan minimizes the differential ways we might discern this shared atmosphere.

Indeed, Brennan advocates for the cultivation of discernment as a strategy against the power of the affects to possess and overtake us, drawing on a longer Western history of affects that were said to grip the soul like demons or other spirits.[13] She claims this awareness of affect transmission and an interactive economy of energies was common into the seventeenth century, finding its "most noteworthy exponent" in Spinoza, but was eclipsed by the egoic affects and individualized emotions of Hobbes and modern psychology, i.e. dispositions one could be said to possess (and not be possessed by).[14] This account of the history of affects cannot be separated from histories of global capitalism and colonialism. In *Exhausting Modernity* (2000) and *Globalization and Its Terrors* (2002), Brennan studied the large-scale acting out of Western paranoia, what she calls "the foundational fantasy," i.e., subject-centeredness, in which the individual takes itself to be a self-contained consumer of the surrounding world, taking up affects from the "broader social order" as though self-originated.[15] The affects that Brennan describes of "Daily Life in the West"—such as grandeur[16] and instant gratification[17]—these are not individualized feelings but sociohistorical tendencies heightened in a capitalist economy. These tendencies are enabled by what Brennan calls "bio-deregulation"—dismantling the "rules" of nature (or at least our limited grasp of these) that govern living things and make possible their replenishment.[18] These capitalist tendencies have material effects on land and air as well as on human and non-human life, including environmental degradation, exhaustion, and stress-related ailments.[19]

Brennan's account of bio-deregulation draws on Marx as it also rests on a critique of his labor theory of value. Brennan writes that Marx fell into the "foundational fantasy" of the self-contained subject as the source of agency and creativity, so rampant in his historical moment, and so was concerned solely with regeneration time for human labor but did not consider the value contributed by (nor the regeneration time needed for) nature.[20] This resonates[21] with Silvia Federici on Marx's neglect of reproductive labor; Marx thought that the world-historical stakes

of emancipation would be played out in the register of wage labor and the production of goods, whereas Brennan and Federici discern how he overlooked the
stakes of reproduction, not only the nurturing of future laborers—roles historically
assigned to women—but the lived conditions in which laborers find sources of
regeneration, the affective and ecological atmospheres through which they move.[22]
As a means of reversing course from the endless expansion of global capitalism,
Brennan offers a principle of benevolent non-interference, the "Prime Directive,"
which she adapts from *Star Trek* (one of her favorite series). For Brennan, this
directive demands that we slow down enough that the land can replenish without
going elsewhere to extract more.[23]

I now turn to the *ought*, to intervention on atmospheres, understood ecologically and affectively through Brennan's Prime Directive. Can one know when the
timing is right or if one's positionality is appropriate to the task of intervening?
As a white person in the US South, can one join in collective acts to intervene on
habits of white supremacy without risking the re-inscription of those historical
dynamics? If the patterning is so thick in the atmosphere itself, how can intentions deviate from the scripts in and through which individuals speak and act and
view the actions of others? To this end I turn to three ways of conceptualizing
the material conditions for atmospheric change: 1) collective agency, 2) the commons, and 3) infrastructure.

First, collective agency is unstable, always at risk of falling apart or, worse yet,
falling back into the institutional inertia from which it first resistantly emerged.
Far from a larger organic super-body, collective agency cannot be neatly mapped
from individual agency. In *Freedom Farmers*, Monica M. White defines "collective
agency" to name the interpersonal dimension of our agency as a separate consciousness—irreducible to individual consciousness—which drives a group of people
united in the belief of their mutual success.[24] Among her practices of collective
agency, I am compelled by "prefigurative politics," to act *as if* desired freedoms
and practices of flourishing are possible. What if we acted *as if* the natural world
was not an infinite supply of resources, always able to bounce back?[25] To build a
community garden is to act *as if* we could produce the means of our sustenance,
as if we are a neighborhood where people care for one another, and to begin to make
it so in the acting out.[26] But who is this "we"? Collective agency as described above
consists of humans, but what of the other species, the land and air that support these
human energies? Drawing on that earlier tradition exemplified by Spinoza and more
recently by Walter Benjamin, Brennan insists that all things are energetically connected, not just individual sentient beings but also the technologies they have made,
as well as the natural world from which the parts were derived.[27] Collective agency,
then, would include all the energies, human and non-human, living and non-living,
dedicated to a shared purpose. A community garden is surely a site of collective
agency; might we describe the garden itself as a collective agent?

Second, a sense of the "commons" might help widen the scope of agency. The
commons name the shared natural and cultural resources of a community, particularly public spaces that are neither home nor work (sometimes called "third

spaces").[28] Lauren Berlant cautions that we must be skeptical of the pastoral, settler colonial history of the commons—e.g. the US context of dispossessing Indigenous peoples and then demarcating a town square—but it is nonetheless valuable for envisioning our more-than-human relationality and the meeting of shared needs, so long as we do not romanticize this uncritically.[29] In other words, when "the commons" presumes a sameness of experience, or that we are all cared for under its auspices, such a concept skates over the difficulty—the inconvenience—of sharing an atmosphere. Indeed, global capitalism functions so smoothly by skating over this difficulty, distancing production from consumption but also speeding up distribution such that one can enjoy one's goods and services with minimal waiting time, far away from the cruel conditions of production.[30] Brennan refuses this complicity with far-away suffering and demands more sustaining arrangements; indeed, she draws on Gandhi in dialogue with Marx to argue for more localized production and the need to slow down.[31] What would be the conditions or scale for a non-exploitative commons? Federici highlights community gardens: "Their significance cannot be overestimated... if we are to regain control over our food production, regenerate our environment and provide for our subsistence. The gardens are far more than a source of food security. They are centers of sociality, knowledge production, cultural and intergenerational exchange."[32] These gardens produce for neighborhood consumption, for subsistence rather than for the market. They recognize the deep connection to land and "make access to land a key terrain of struggle."[33] Land is a site of struggle here not for ownership but for stewardship, for creating something in common, collective bonds, which Federici argues we do when we engage in "commoning" the material means of reproduction.[34]

Third, and another way of describing the work of commoning, "infrastructure" names the material conditions for relational life, the "below" structure that enables the operations of the whole. This includes transportation, communication, the management of water and waste, and related systems that, when functioning, fade from concern into the background.[35] Much of our infrastructure is inherited and can be hard to see in its familiarity, even harder to imagine changing, but it has been the work of activists to push for infrastructural transformation toward collective flourishing,[36] whether through curb cuts or cleaner water or access to fresh vegetables. A community garden is an infrastructure, an atmospheric intervention.

The certainty that our actions will benefit more than they harm cannot be had, but then again white people have been especially dangerous when most assured of doing good. Deepening this ambivalence, the demand of Teresa Brennan's Prime Directive might be clear, but how exactly we develop these habits of restoring—of slowing down so that nature can replenish—is far from cut and dry. Who gets to build the community garden, where, and who sets the rules? Here, we might extend Linda Martín Alcoff's four interrogatory practices in "The Problem of Speaking for Others": 1) examine our motives to intervene and 2) the bearing of our location on what we can perceive; 3) remain open to criticism; and 4) attend to the probable or actual effects of the intervention rather than the purity of intentions.[37] When speaking for an atmosphere (no less than speaking for others), error is inevitable; but speak we must, and slow down we must, as Brennan bemoaned twenty years ago, before global capitalism runs its extractive course.

Notes

1 Teresa Brennan (1952–2003) grew up in Melbourne and Sydney. She pursued BA and MA degrees alongside activist initiatives before going to Cambridge in her thirties. Brennan's dissertation became her first book on Freud and femininity. She would go on to publish four books (a fifth posthumously) and two edited volumes, all in the last decade of her life. In the 1990s, Brennan developed an innovative albeit short-lived doctorate for Public Intellectuals at Florida Atlantic University. In late 2002, at age fifty, Brennan was struck in a hit-and-run from which she eventually died.

2 The term is credited to geographer Ben Anderson, who defines an "affective atmosphere" as a collective, pre-personal or transpersonal sensibility. Anderson, "Affective Atmospheres," *Emotion, Space, and Society* 2 (2009): 77–81.

3 Kristen Simmons, "Settler Atmospherics." *Fieldsights* (2017): https://culanth.org/fieldsights/settler-atmospherics

4 Eric Stanley, *Atmospheres of Violence: Structuring Antagonism and the Trans/Queer UnGovernable* (Durham: Duke University Press [UP], 2021); see also "The Affective Commons," *GLQ* 24, no. 4 (2018).

5 Sara Ahmed, *The Promise of Happiness* (Durham, NC: Duke UP, 2010); see also Ahmed, "Atmospheric Walls": https://feministkilljoys.com/2014/09/15/atmospheric-walls/; Jack Leff, "Expanding Feminist Affective Atmospheres," *Emotion, Space, & Society* 41 (2021).

6 Judith Butler, *What World Is This? A Pandemic Phenomenology* (New York: Columbia UP, 2022), 12.

7 Harriett Washington's 2020 book is a helpful introduction, as is the work of Robert D. Bullard, who shows that race, rather than class, is the primary factor of risk for toxic dumping; NIMBY sentiments do not adequately grapple with another acronym, Locally Unwanted Land Uses (Washington, *A Terrible Thing to Waste*, Little, Brown Spark, 117).

8 Jean-Thomas Tremblay, *Breathing Aesthetics* (Durham, NC: Duke UP, 2022).

9 Lauren Berlant, *On the Inconvenience of Other People* (Durham, NC: Duke UP, 2022), 19.

10 Teresa Brennan, *The Transmission of Affect* (Ithaca, NY: Cornell UP, 2004). I elaborate this response to Brennan's critics elsewhere: Guilmette, "Unsettling the Coloniality of the Affects," *philoSOPHIA* 9.1 (2019), 73–91.

11 Sara Ahmed, "Happy Objects," *The Affect Theory Reader*, ed. Gregg & Seigworth (Durham: Duke UP, 2010), 37.

12 Berlant 2022, 19.

13 Brennan 2004, 97–98.

14 Brennan, *Exhausting Modernity: Grounds for a New Economy* (London: Routledge, 2000), 12–13, 41. She continues, drawing on French feminist Le Doeuff, that "Hobbes and Shakespeare agree in what they deny: that fancy can pass from one person to another. Each individual becomes a closed space of relation to their fantasmagoria" (2000, 44).

15 Brennan 2000, 8.

16 Brennan 2004, 105. *Grandeur* names the closure of a critical gap between one's own interpretation and reality as such; this certainty may justify aggression, righteous indignation, and other projections (106).

17 Brennan 2000, 156.

18 Brennan 2002, 30.

19 Brennan 2002, 19.

20 Brennan 2000, 11.

21 Brennan and Federici do not appear to have been interlocutors, so the resonance might bear witness that these ideas are "in the air," so to speak. Silvia Federici, *Revolution at Point Zero* (Oakland, CA: PM Press, 2020).

22 Federici writes, Marx "remained wedded to a technologistic concept of revolution, where freedom comes through the machine," and that he thus adopted the capitalist criteria for work as waged and contractual (Federici 2020, 105).

23 The Prime Directive for Brennan seeks to "reverse the concentration of wealth on the one hand, and the distance over which natural resources can be obtained on the other" (Brennan 2002, 156).

24 Monica M. White, *Freedom Farmers* (Chapel Hill: UNC Press, 2018), 7.

25 On this point and others, George Washington Carver was far ahead of his time (White 2018, 46–49).

26 We built the garden to construct more nurturing and sustainable relations. First we cleared an abandoned lot, and we heard from older neighbors about a house that burned down; we could discern the grass dipping unevenly into the shape of its foundation. We learned from neighborhood kids about the pet cemetery between the black walnut trees, midway to the creek. We asked the advice of a soil scientist and a farmer where it was safe to till and where it was better to build raised beds from corrugated panels. The hope was to disrupt toxic energetic circuits—to grow fresh fruits and vegetables, to construct tire swings, to collect a library of books and toys—to change the atmosphere.

27 Brennan 2000, 43.

28 Joanne Dolley, "Community Gardens as Third Places," *Geographical Research* 58, no. 2 (2020): 141–153.

29 Berlant (2022) follows Federici in arguing that "the attachment to the common… too often stands as an aspiration to consensus that tries to make affectively simple the nonsovereign relation" (80).

30 Brennan 2002, 10–11.

31 Brennan 2002, 14, 152–156.

32 Federici 2020, 159; Federici observes that urban gardens started "thanks mostly to the initiatives of immigrant communities from Africa, the Caribbean or the South of the United States" (ibid.). See also: Braga Bizarria et al., "Community Gardens as Feminist Spaces," *Geography Compass* 16, no. 2 (2022): e12608.

33 Federici 2020, 160.

34 Federici 2020, 162.

35 Ara Wilson, "The Infrastructure of Intimacy," *Signs: Journal of Women in Culture and Society* 41.2 (2016), 274.

36 For a beautiful example of this, see *Crip Camp*, dir. Nicole Newnham and James Lebrecht (*Netflix*, 2020).

37 Linda Martín Alcoff, "The Problem of Speaking for Others," *Cultural Critique* 20 (1991), 5–32, esp. 24–26.

Chapter 16

Listening to the Earth

Wolf and Lisa Wahpepah in conversation with Michael Kearney

Michael Kearney: Let me begin by citing the mission statement of the Hosting Earth Project:

"The Hosting Earth Project engages the question of ecological hospitality by asking what it means to be guests of the Earth, as well as hosts. In our current climate emergency, it challenges the anthropocentrism of Western culture which celebrates the sovereign human subject as "master and possessor of nature"—Descartes. But what if we experience the natural world as a place which holds and hosts us: animals, trees, plants, birds, mountains, rivers, and seas. What if we see Brother Sun and Sister Moon as nourishing us with their cosmic vitality. In short what happens if we reverse the conventional model of hospitality and acknowledge the double sensation of hosting/hosted, a reciprocity principal which signals an ecological interspecies connection between all living beings. The Guestbook Project of Hosting Earth is committed to the pursuit of climate justice as we face the environmental emergency of our time"

My first question is if there's anything that you would like to say in response to this mission statement of the Hosting Earth Project?

Wolf Wahpepah: It pleases me to hear the notion of *hosting*, because of course from the Native perspective, it's Mother Earth who is the host and we are the organism, one of the many organisms reliant upon the host. In Native culture, we use the phrase "For All My Relations," which implies an interconnectedness of all life. If you push a domino anywhere in the chain, it affects the rest of the chain from that point forward. And so, our culture is known for having been here for thousands of years without significantly changing the environment. Of course, we see all living beings as entities. The world that Native culture lives in is a world that is filled with life. If you look at this world

DOI: 10.4324/9781003456940-21

through dead eyes, you'll see it as inanimate and dead, but if you look at this world through eyes filled with life, and that's what you're looking for, that's what you'll see.

Everyone refers to "troubled times." You know, from a Native point of view, it's been nothing but troubled times for a little over 600 years. But each generation, in any culture, will tell you about the troubled times that they went through. Today we're noticing heightened troubles at a mass level. Between climate change and unstable politics. The people of the Earth are reliant upon our mother, the Earth, who has sustained us all this time.

When you look at the supply line breakdowns we're noticing globally, it's amazing what the smallest thing can do to throw off the whole system.

Well, the complexity of the economy, even given humanity's diversity, is but a mere fraction of the complexity of nature. Nature is far more intricate and sophisticated and complex than any man-made system. And so, it's even more vulnerable to changes. But it also has the ability to recuperate and to adjust, because it's alive and everything that's alive has the ability to evolve. And so, we in these modern times find ourselves in a system of interdependency where we've tipped the scale on several of these systems and cannot stop the cascading effect that's manifesting.

And for what it is worth, all of these things were prophesized when Europeans first found Turtle Island. Our living elders at the time watched as the Europeans created settlements and we looked at how they divided labors. We would watch them from the outskirts of their dwelling places. We would observe them when they woke up in the morning and compare the first thing they would do to the first thing we would do. We would watch how their whole day really didn't make sense to us. We would watch how they would rely on resources that they couldn't produce. Our elders saw that this was a short-term thing, that this approach was not sustainable and that it wouldn't be possible to endure through the cycles of time. And so, 500 years ago or longer, we already had elders saying, "this is doomed to fail."

So, from a Native perspective what's happening now is merely what we expected to happen. We understood that if we tip the scales too much we cause damage that takes a great deal of time to reverse. And we're aware that it's possible to go too far and to make things irreversible. So based on our simple yet actually more complex view of the world, we decided that we should be cautious about changing it drastically because of unseen consequences that could happen.

Another precept of Native Culture is what we call, "honoring the generations." Honoring the generations has a double meaning for us. We say that when making an important decision, we should reflect on the previous generations and the wisdom that we gleamed from them, the experiences that they had and the insights that they have gained and the instructions that they've left us with. That we should honor those individuals for what they went through to provide the future generations with this knowledge. And, at the same time, we ask ourselves when making an important decision how this might affect the seventh unborn generation from here? If we care about the future generations, we have to consider them before they arrive.

I do like the fact that people in the Hosting Earth Project are gathering to discuss these issues and I realize the people who will see this conversation are already on our side. So, even though this is like preaching to the choir, I hope that we can provide an insight or two, or at least some inspiration. Because it's all hands on deck and with even one more person being concerned about this, that's the only way we're going to get there. That's why we're having this conversation.

Lisa Wahpepah: Yes, I think the Hosting Earth Project is a wonderful thing, in particular because it's something that's going to be used as a teaching tool and that it's something that will continue to be used.

MK: Could you tell us what is meant by the concept "For all my relations"?

LW: It's realizing that we're related to everything, which is of course the Earth. The Earth is our mother. Our mothers are life givers, life bringers, and they nurture us, and they give us everything we need to live. They feed us, not just physically but also mentally and emotionally. All living beings are our relations—not just two legged but also the four-leggeds, the winged, the insects, the finned that live in the ocean, the trees, the plankton, the flowers. It is everything—for *all* my relations. So, that's my understanding of what we're speaking about when we speak of *all my relations. All* of it.

WW: A lot of our teachings are based on an intrinsic understanding that isn't always depicted in our words. For example, we realize that we relied on the four-leggeds to feed our people and yet we regard them as a nation. And so, when we were preparing to go out hunting for example, we might crawl into the sweat lodge, both to de-scent our bodies but also to pray; and while we're praying we might sing the four-legged song. The meaning of the song isn't actually revealed by the few words in the song. The attitude that underlies it is that we're singing to the nation that we're relying upon to feed our children and we have nothing but regard for them because we want them to be able to continue to feed us; and so we might sing the four-legged song to get their attention and to let them know we're coming, but with a certain attitude that would be agreeable to them. The attitude that we would hunt with is like this: "we already know that some of you aren't going to survive next winter and we're praying that those are the ones we come upon so that we don't harm your nation to feed ours."

This was just an understanding, it's not revealed in the words, it was just understood that by approaching it this way, our stories say, the game would just present itself—that they understood the hunter–prey relationship and they understood the role of the original instructions that the Creator gave their nation. And when the two leggeds embrace the same instructions that the Creator had given, well, what we find is a natural world that actually cooperates with us. That we don't have to master or dominate or control. We just have to cooperate with it. There is a vast difference in attitudinal approach. And by learning to cooperate we can have a blessed life and not at the expense of other living beings, without destroying or damaging what can only be referred to as the natural order. And preserving the natural order is how we will preserve humanity.

LW: Learning, or re-learning, to live in harmony and honoring all life. Honoring all life—that's the important part.

MK: You mentioned the phrase, "original instructions," and how everything in the natural world lives in its original instructions but that we two leggeds, at least a lot of us, don't. Did I understand you to be saying that if we come back into living in accord with our original instructions, living in harmony with the rest of nature, that the natural world in some way recognizes this and responds to us differently?

LW: This is an old conversation really, yet, somehow, in recent years, it's reemerging as a new conversation. It's as though with the industrial revolution we forgot about the intelligence of the Earth. It's about remembering to learn from the Earth herself and the immense intelligence that exists within the Earth—the intelligence of the trees and the plants and the soil, and all of it. I think that's the part that we two leggeds need to realize and remember—the intelligence, the immense intelligence of the Earth which brings us back to our original instructions and helps us gain a higher level of intelligence.

WW: We've evolved as a species to adapt to unnatural conditions and to a certain extent we've been able to do that because as two leggeds we're very adaptable. Turns out that life is very adaptable. But if we can be adaptable to what I'm going to call forms of corruption, well, we're even more adaptable to the laws of nature. We can revert back to natural conditions even faster than we accommodated to unnatural conditions. We're designed to respond to natural stimuli. If we're provided with natural stimuli, we just engage.

The problem is that our lives have become so unnatural. We change hours on the clock when a season comes. We're going to pretend an hour moved. The sun didn't move but the clock did, you see. Things might be in the same place, but we call them something different. It's very confusing to the mind. If you confuse the mind enough with variables such as these, pretty soon it's almost impossible to notice the unnatural effects that are ensuing. They just kind of happen subliminally without us even noticing them and soon become normal causes as we've adapted to that abnormal condition.

Then someone has a baby, and the baby is born into that unnatural condition. From the baby's point of view as the baby grows, the unnatural condition is normal and has been normal their whole life and only the elders can see the difference, which is why we always had councils of elders to advise the leaders. And those councils of elders had more life experience and they themselves had benefited from the previous generation's elders. So, they had been clocking patterns and developments in the natural world and they could see when things were shifting in a certain direction.

We don't have to read that climate change is happening, we literally just have to step out our front door or look out the window. We've watched several species of birds move into our region that never lived here before. In the past five years we've seen three species of birds that have been displaced from somewhere and now they're here. We've watched so many firsthand changes, and for several years now, the complete absence of water. The fact that our stream, which was a seasonal stream in the first place hasn't flowed in almost three years—that never happened before. We've been on this mountain for a total of some thirty years, and we've become aware of the normal patterns of how winter should go and how summer should come. And again, you really don't need a scientist to tell you that the world is changing, unless you're living under fluorescent lights 90 percent of the time and you aren't getting any input from the natural world because you're so separated from it. Then, the only way you would know is if a scientist told you.

MK: Could you say a little more about the principle of reciprocity and why it's so important?

WW: Something that I think a lot of people don't realize about our traditional ceremonies is that every time we conduct a ceremony, energy that we've been carrying inside ourselves is transferred, is gifted, is returned back into the Earth. This energy is pure life force. And so, the original people, they understood that while the Earth is vast, it's not infinite. Vast means big but still finite. And so even the ancient ones understood that we lived within a finite system, which means that we can draw more from it than it can recoup. Think of it like a checking account, the only way we can write more checks is by making more deposits. Our ceremonies understood that.

One of the most fundamental aspects of Sundance is that it is a renewal ceremony—among other things. The purpose of Sundance is to support the procreation of life and so the dancers give of themselves by fasting from food and water so that their life force at the altar can be transferred back into the Earth. This is so that the next generation can draw on the same "bank account"—so that they will inherit the Earth.

Reciprocity is irrevocably essential to the perpetuity of life. If we don't keep putting life force back in we can't keep pulling the life force back out. There were fewer of us when we started this conversation today than there are right now. Already there are more of us, by tomorrow even more. The more of us, the more consumption, the more the imbalance is perpetuated. Unless we actually rework our systems from top to bottom. But these are changes that everyone is reluctant to take. It sounds good idealistically until we say, "To make these adaptations there will have to be some limitations you don't currently live." Well, all of a sudden, all bets are off. We watched people protesting during COVID because they couldn't get a haircut.

LW: We moved from a reality of "you take what you need" to "we take far more than we need," and it's on a systemic level of growth that cannot be sustained. It's too much. You know living in balance is what kept the landscape pristine for thousands and thousands of years. Living in balance—not taking more than you need. Western civilization has made us weak and in that weakness we think we need more and more and more. That's what's cost us and we're now living in severe imbalance.

MK:	What is at the root of this weakness, do you think?
WW:	If you expect life to provide more than it was designed for, you'll never be satisfied. And a lot of people have inflated expectations from life that are far beyond what they need to be happy and content. When these expectations exceed what is realistic, what you have are people living in a permanent state of discontent and no matter what they have they feel like it's not enough, so they're not satisfied by what they've already received even if it's more than what an ordinary person ever received 500 years ago. It's hard for most modern first world countries to realize that we're living better, that our standard of living is above that of kings and queens of the Middle Ages. We're eating fruits out of season from foreign lands. We're traveling hundreds if not thousands of miles on a regular basis. An ordinary person has seen more territory than kings and queens used to see and yet it's not enough.
MK:	Do you think that at the root of this discontent is our disconnection from the Earth?
LW:	Yes, it's about a deep disconnection from the Earth. When we understand our symbiotic relationship with the Earth we are in touch with what the Earth gives to us. When you think of your mother; what was it like for you as a child, as a teenager, as a young adult, as an older adult? When you think of what she gave to you and how she nurtured and cared for you. But our relationship to the Earth, that's the part that has been disconnected. She's not seen as our mother who provides for us and nurtures us. So, that deep disconnect, you can't see it from where the mind is.
WW:	If you divorce yourself from the innate satisfaction that the spirituality of the natural world has to offer, all that is left is self-gratification. And because we can over-produce, we can take plants and turn them into concentrated sugar hundreds of times more potent than how it could be found in the natural world. When we realized that we could do things like that it became about how much more gratification we could get. But the more gratification we focused on, the less satisfaction we had. Self-gratification is never enough because we two leggeds always notice the part that we don't get, "I missed a spot... the one that got away."
MK:	Is this why the attitude of gratitude is so important?
WW:	Gratitude puts us back in proper relationship. Gratitude is a big conversation. People have been told—with the shaking finger—"you should be grateful!" and that doesn't work. We would say it differently. We would say "you could be grateful." It turns out that grateful people tend to be better adjusted

and happier than ungrateful people, so it really is in each of our own best interests to learn gratitude. You'll have a happier life. Instead of complaining about what you didn't get, notice what you have. And ask yourself simple questions such as, "how many people on Earth don't have one of these?"

You've known us a long time Michael. The first five years we lived here in this camp we didn't have a refrigerator. When we left camp we had to bring our meat with us in ice chests so it wouldn't go bad. And then we got a fridge. The community took up a collection and bought us a propane powered refrigerator—we didn't even know there was such a thing. So, in our remote wilderness location we can power a propane refrigerator. Then we read that at least one-third of the world's population in India and other Third World countries doesn't have one.

We would drive down to town for ceremony and I'd be down there in the sweat lodge and the thought would cross my mind that I have a cold drink waiting for me in my fridge at home and just the thought of that made me feel good. And I would look forward to the cold drink waiting for me. I know that almost everyone has a cold drink at their fingertips so much so that they couldn't possibly get the same degree of satisfaction from it.

The mind becomes conditioned to see things in a patterned way and once the pattern has been established it takes some work to break that pattern; but there are moments of insight, of realization that can occur. With our traditions these epiphanies, as it were, frequently happen in ceremony and since we spend a considerable portion of each year in ceremony, these insights would find us just naturally. I guess what I'm saying is once an unnatural pattern has become ingrained it's going to take work to break that pattern because the mind is going to keep wanting to follow the same path.

But can it be altered? Yes, it can be altered unnaturally which has caused so much social disease, depression, suicide, addiction, the need for anger management courses. There's so many different kinds of upset going on right now but these things can be rectified by the same thing we talked about earlier—by returning to natural conditions, by having a relationship with the sun and the moon.

I remember I was teaching someone how to drive in Los Angeles. She'd lived there all her life. We were in the car for a short time, and she asked me if I was Native. I told her yes, and then she looked at me and it just threw me off, she looked at me and asked me if I'd ever seen a full moon. I thought it was a trick question. I didn't answer immediately, I was trying to figure out what was behind it. The best I could come up with was, "yes" and then she immediately said, "me too," and now I really was confused. I was thinking, "is she talking about that one time she saw the moon?" OK, well, once I had been in LA for a while I realized that all the buildings were tall, and you don't really notice the sky and the lights are so bright out down here that you can't really see the stars. Could it be that she was this far removed from the natural world by living in Los Angeles that she was enthusiastic about that one time she saw a full moon?

You have to consider who you're talking to, but we would say, "Crawl in some sweat lodges. Go up on the hill for Vision Quest, spend four days all by yourself in the natural world without food or water inside of an altar with ceremony to put you there. Do these things and watch how the unnatural patterns in your mind disappear quickly, surprisingly fast. Within a day you're thinking has already changed. Things that normally cycle through your mind every twenty minutes aren't doing it now without the stimulation being there and your mind focuses instead on natural events again; as soon as you interact with what you were designed to interface with, it has an immediate effect."

I mean people will go up on the hill one time and come down and see the world differently. You could've written down the instructions and it would be cerebral and in their head. You could make a six-month course, it would be a whole semester, or you could go up on the hill for four days and you would just come back with it.

You asked about gratitude. It was not uncommon with many tribes that when a child approached puberty one of their elders, maybe their grandfather or their uncle, would walk them out of the village and set up a vision quest altar and tell them to wait here until they came back and then come back in four days. These could be 12-, 13-, 14-year-olds. Once they found themselves no longer within the safety and comforts of the village, they realized what their life would be like if they were all alone and they would sleep on the hard ground every night. They'd understand now what cold is because they couldn't have a fire. They'd get a new relationship to thirst because all their life just like everyone when they get a little thirsty they just take a drink. But now, up on the hill, they didn't have it.

What I'm trying to get at is that this is a non-intellectual way of giving instructions that penetrated deeper. When they came back you didn't have to tell them to respect water, they just did, having spent days without it. Now they realized how much they appreciated their family and their relatives after four days alone with no one to talk to. Now they had some perspective on how life could be. That perspective is essential. Without seeing the difference between this and that, we don't know what to be grateful for. In fact, unfortunately, what happens is the comforts that we've always been provided, we come to think we're actually entitled to and that we have a basis for being disgruntled if we're denied them—even though that was a bonus in the first place.

So, you know, a lot of cultures talk about the hungry spirit—no matter what you give it is not enough. That is what Western Civilization has become—a ravenous hungry spirit. We glorify the super-wealthy. They make TV shows about themselves that poor people watch. They have presidential inaugurations where this one spends even more than the last one spent and we somehow glory in their lavish luxury.

In Native cultures, like in the Iroquois Confederacy, which I'm familiar with from my tribal background, our tribal leaders weren't allowed to own anything. They weren't allowed to have possessions They weren't glorified and made like kings and queens. They lived at the same level as everyone else, and, because of that, they couldn't be bribed. They couldn't be co-opted. Lobbyists simply would be out of work. They'd be in the welfare line.

The Founding Fathers of the United States copied the Iroquois Confederacy's tenets, but they were property owners and when they got to the part about giving up their property, they'd say, "Well let's just skip that part." And that part was critical. It prevented exploitation in the future. Again, we see it now. I forget the numbers—how many lobbyists per congress person in Washington, DC, something like 12 to 1.

MK: You were talking about how our disconnection from nature is at the root of a lot of our suffering and the suffering we then inflict on our world. You were talking, Wolf, about how ceremony can bring us back into that kind of harmony and that different perspective.

I really feel very strongly that we non-Natives need to listen deeply to the wisdom of teachers like yourselves. But what about those who don't have the opportunity to come to ceremony? Obviously you'd encourage anybody that could do so to find Native teachers in their part of the world, but what if that is not possible?

Are there other ways of coming back into connection with the Earth?

WW: Yes, there are. Start making friends with life. See life as living, not as inanimate.

When we made this spirit camp, we understood how the natural spirits of the natural world can benefit the two-leggeds tremendously. So, you know, a lot of Western Civilization tries to set up something like a farm and now they have this problem, they call them "varmints." Now each of those is a living organism and there are different species, but they are all just varmints. That's a way of looking, a way of thinking.

When we came here global warming had not advanced to the state that it is now. California has always had droughts. When we first came here it was a drought but it was a normal drought. We knew that we wanted to make friends here and that we were the newcomers to this mountain. So, the first thing we did was to identify how many of the pine trees on this mountain were actually in our camp. We realized that they were thirsty. We had no irrigation system at all. We had the well, but we didn't even have a water tank.

One of the very first things we did was over a period of weeks to carry 5 gallon water jugs to each and every living pine tree in our camp. They hadn't had a drink in quite some time. Here in this camp, the summers have only gotten hotter and hotter. The deer population struggles for water, in fact, all of the various four-leggeds do. We put out water bins and water basins throughout our camp and they came and drank. The yellow jackets came and drank. To this day, they find the water and they'll spend all day taking microscopic sips—thousands of them, and we'll watch the water level in a bucket go down.

We're giving it away, in what we call "the spirit of give-away." A lot of people have come to finally recognize what we have said for a long time. Even science is saying it now. We have long said "water is life." So, in the spirit of give-away, we give away life—we offer water. We put it here and there. A lot of people don't like to attract yellow jackets, but we have made friends with them. And people come to our camp, no one gets stung. No one gets bitten. Because they see us as friends. What I'm saying isn't just a way of thinking. There is a consciousness involved that permeates the natural world and when you regard its consciousness with your awareness, it feels validated. And as soon as it's recognized, like when someone knows you by name, you can make friends with them. And the more you make friends with them, the more they'll accommodate you to an extent that your requests are not actually unreasonable. And that is where humanity has gone wrong. We've made too many requests that are unreasonable. Now we are getting negative reactions from the natural world.

LW: I'm thinking about this conversation that we're having. I think part of the problem is fear versus respect. When I was a little girl and we were going harvesting and gathering from the natural world, we were taught to respect. You respect nature and all living beings. Whereas, when I think of stories like *Red Riding Hood* or *The Three Little Pigs*, or what have you, nature has become something to fear and that is sort of what's taught. I think this has allowed for the great disconnect that we're seeing everywhere. Whereas in Native cultures it's taught differently. We're taught respect, a healthy respect. That there are things much greater than us and that the two legged aren't really the most powerful living beings on the planet. What's happening with the climate emergency is that the Earth is letting us know that healthy respect is not being shown. We're not living in harmony with the living beings that provide our entire existence.

MK: How do you think the Earth regards us two leggeds?

LW: That we're out of balance.

WW: We have become parasitic. We threaten the host, the natural world. The Earth regards all life and wants to give life. The Earth is not a destroyer. The Earth is a Creator. We are the ones who have thrown the life cycle out of balance to such an extent that now we threaten the organism that sustains all life. At some point in time Mother Earth has to decide how much can be salvaged and preserved because she has to make sure that life itself is not extinguished. If she has to make decisions or cause consequences that make it less habitable for us in order to diminish our negative impact because of our numbers and our practices on the creation, she will. Any mother would.

It is better to think of the Earth as a mother, as opposed to objectifying her. Just think about it. The Earth is not *like* our mother, she *is* our mother. And how you would care for your mother is how you should care for the Earth. And even if you're part of the species that is running rampant and causing trouble, you as an individual can still make friends with the Earth and show regard.

We made friends with the ravens. They are spooked by almost everyone who comes to our camp, but we made friends with them, so they are not spooked by us. Even if our species is making grandiose mistakes at a mass level, any individual can stop objectifying the natural world, start giving it life, and watch it give life back. The moment you do, it does. We are the ones closing ourselves off from the blessing. Mother Earth is nothing but giving. Her hand is reaching out all the time, always, for us.

Make friends. A lot of people might say, "Yes, but how do I do that?" Well, give offerings. Here's a few basic examples: give water to the trees; spend time in the natural world, observe the changes; put something out for the creatures; pray about it. Don't throw food at them! Offer it, offer it and say: "that's for you"—in the spirit of give-away, no expectations, no strings attached.

LW: Don't take more than you need.

WW: And thereby discover just how generous the natural world is. You can tap someone's generosity by taking too much, by always asking for more. We say people sometimes deny the helping spirits. We say that the helping spirits often are spirits of our own deceased ancestors and that they intervene in our human affairs from time to time as they have permission. Sometimes we, as two leggeds, take them for granted. We forget where the center of the universe is. We forget that life came from the spirit world. We focus on our immediate circumstances and our immediate appetites, but this is at our own expense because it leads to the downwind of dissatisfaction, depression, and all kinds of other abnormal spiritual conditions.

MK: If our mother, Earth, were to talk to us two-leggeds, what would she want us to hear right now?

WW: She's crying. She's crying. The unnecessary devastation. All these forms of life are her children. How does a mother feel towards her children? She watches how some of her children are killing other of her children. What's a mother to do?

LW: It's the immense lack of regard. You know, what would mother say? "Do you not see that you are me?" I mean even on a biological level we are the Earth. So, "Do you not see that you are me?"

MK: While we are having this conversation, the most awful war is unfolding in Ukraine. Wolf, I've heard you telling the story of The Peacemaker. You were talking earlier about the Iroquois Confederacy. I just wonder if you could share some thoughts about what we can learn from the story of The Peacemaker. What can we hear in that story that's important at this moment?

WW: Well, the man that we call the Great Peacemaker who was alive when the Europeans started first coming, he'd heard about them, but he had never met one. A lot of Indian people were aware that a new person had arrived on Turtle Island. The Great Peacemaker was given a vision by the Creator because he had tremendous empathy for the people. While he was praying, the Creator showed him that of all the forms of suffering in the human condition exactly one is optional. We can't do anything about aging and death. We can't prevent injuries and we can't completely prevent illness. But this thing that we do, this violence that we impose upon one another has become a significant aspect of human suffering and it is the only one that is optional. If no one did it, it wouldn't happen.

And so, the Great Peacemaker was shown a vision of the Tree of Peace where he saw that the diversity of the people could still be unified. And even though there were sub-groups and many tribes, the Creator showed him that they could form a union like a tree growing with different branches. And these different branches would be the different nations of people that lived, all joined by one trunk, and that if they were in proper relationship to one another in a system that wasn't vulnerable to corruption, that it would be a system based on fairness. By incorporating and adopting more and more tribes, his intention was to literally end all warfare on the North American Continent, on Turtle Island. Because everyone knows, Indian wise anyway, that no-one would ever attack a member of their own tribe. There was a time when this was absolutely true. Now in the modern time we've incorporated by virtue of assimilation many of the disorders of the greater society. So, you'll find exceptions to these things now even in Native life. But there was a time when no one would consider attacking a member of their own tribe. So, if you can get all the tribes to join as one, there is no longer an enemy to attack.

Some of the words of The Great Peacemaker recorded in wampum belts indicate that he was trying to prevent more crying women, more mothers and wives losing their son or their husband to an enemy attack.

How does this apply to this conversation? I'll say there is some overlap that I can see in that he was prepared to regard all people as the Creator's children. Throughout most of his lifetime, there had only been one race here. Now he heard about another race arriving, although he never saw one. But the Creator gave him a vision about them that showed him far into the future, into a time that really could only be compared to now, a time when the Earth was in such unrest that life was no longer sustainable for Native culture. And what he saw was that some of the white people were like-minded with the Native perspective on life. And in his vision, and in their society, and in the Iroquois Confederacy they were prepared to receive all people on an equal basis.

So, there was the time, in the mid-1550s, when Europeans themselves were not content with their own societies, which is why they left Europe and when they got here, they weren't always content with the settlements they became a member of either. Non-Indians, meaning in this case white people, were welcomed, in the Great Peacemaker's own words with the same love that one would have for a long lost brother. That is how it was spoken and is remembered.

I only think to connect those dots by saying—learn to make friends with everything. I know that this has become another cliché. Some people say it's unrealistic to think that you can make friends with everyone. That is so, but in the effort of making friends, you will make friends with at least some of them. So, when you see a tree, a wolf, another person, instead of thinking varmint, or enemy, or nuisance, we could see the other as a friend. If what you saw was the possibility of a friendship that hasn't started yet your interaction would go very differently. If you approach them guarded and with fear they're going to think there's a reason to be guarded and react with fear, and they're going to match you.

So much of what we inherit is what we give birth to, and we think it came from outside of us, but we initiated the dynamic that set the tone for the events that followed. And this objectifying the world, we blame it on the object, "What just happened?" And we'll isolate an action that they took, "But they did this!" Yeah, but what happened before that? What was the first part? Because how it started affects the conclusion. I think what goes wrong is what I referred to earlier—our preconditioned negative thinking really gets in the way.

MK: Our mother, Earth, does not have implicit negative bias. If we come into a deeper connection and harmony with her, can she re-educate us?

LW: Education is always available. It's whether or not people have the wherewithal. But, it has to start young. If all a child knows are cement sidewalks, the power grid and sterile environments, they don't even have an opportunity to make the connection. That's a huge part of it. This has been going on for quite a while now. It seems to be getting worse for the next generation to live with that connection. A good place to start is to stop depicting indigenous cultures around the world as savage or ignorant. It's time to start tapping into the immense intelligence of indigenous people who are living in harmony and connection with the Earth, and to honor and respect the life that gives us life.

Honoring and respecting means understanding that everything must die for us to live. It goes all the way down the chain like that, so it is very much a symbiotic relationship. But Western society, Western civilization, modernization - there is such a disconnect. Where do all these things come from, where does all this stuff that we have, such as our food, where is it all coming from?

We have to start relearning. But how do you do this when industry controls virtually everything? And if industry doesn't make changes then the people will continue to get their needs met by what's provided. If it was provided differently people would get their needs met differently. It starts with that in my opinion.

We have to relearn healthy respect and regard from indigenous cultures around the world and someway somehow start demanding that industry must change.

WW: They did a study on genetically modified food and, because it's alive, it has a DNA, and it has an unnatural DNA because it is the by-product of combining aspects of more than one species into a life form. And so, they showed that as people consumed more and more genetically modified food, it actually altered the genetic code of their own DNA, it modifies chromosomes inside of them because of its DNA. That leads down a path towards untold abnormalities and possible illnesses. So, they did another study the same group that they had been studying, taking them off of genetically modified foods and putting them on only organic food. And their DNA reverted back to its original instructions in a much shorter period than it took for it to become altered. The world we live in is designed for us to flourish naturally and this happens as soon as we remember or relearn our right relationship with the natural world as a species, because some of the individuals never learned it in the first place. Which is why Lisa refers to the need for education. The more we reclaim our natural behaviors and come into proper relationship to the people in our lives, to the Earth, to the Great Mystery that is the origin of all life, the healthier and happier we will be. The biggest question that even Stephen Hawking cannot answer is, "how did the first particle come to be?" When we come back into proper relationship with these mysteries and realities, we become whole again. All of a sudden we don't feel like we're not enough. But right now we have holes in ourselves. We don't feel right. We take a pill. We entertain. We find all kinds of hobbies to distract ourselves.

We actually *are* the Earth. We're made out of the metals, minerals, and liquids that are in the Earth walking around in ambulatory form. By virtue of being the Earth we can never in reality break our connection to her. When we sense that she's in distress, it becomes a sense of distress within us so that our mind will identify the source. The Earth is talking to us so that we'll engage in the conversation, because there is something that we can do about it. From a psychologist's point of view, the patient will say: 'I have a sense of impending doom, but I can't link it to anything obvious in my immediate environment. So, they'll say, "Oh, that is an irrational concern, maybe we'll call that paranoia."

We have a name for anything that affects our performance and our productivity. We have a name for that and usually we have a pill or a course of therapy that involves a professional. A lot of people don't realize that the reason they're on Prozac etc., is only because they're a deeply spiritual person and their connection to the Earth is still intact and they can sense a form of distress; but they can't link it to anything obvious and they don't know what to do so they become paralyzed or confused; they get labeled with some name or put in a box with some category.

But really and truly, the people of the Earth are sensing what's coming—that we are in an extinction event. You know people want to focus on positive things and be happy all the time. There is just no getting out of it. There's something depressing about extinction events. Everything is dying around us. Species are disappearing, for real. They can't assure us that when my granddaughter grows up she'll get to see a giraffe. The world that we knew is now in encyclopedias or in pictures online. Eventually, we're going to show kids pictures of giraffes and they're not going to believe that it wasn't photo-shopped!

So, to keep trying to answer the most difficult question: for everyone who isn't already connected to a form of spiritual practice or discipline in some way, what can an ordinary person do on their own?

Start somewhere. Start going for walks. Go for walks in places where you don't have to fear other two leggeds so that fear isn't part of your walk. Find places to walk that are relatively safe. Make friends with the plants and the animals. Maybe sit in one spot on a regular basis because you'll notice changes by having a fixed location, and you'll notice variables change. This is how we anticipate the coming of the seasons. This is how we can tell. See, Winter doesn't start every year on

December 21st. Sometimes it gets cold and snowy before that. You can't really base your life on the calendar, you have to base it on the environment and observing the climactic changes happening around you. An animal knows that it's getting colder, so it starts growing a thicker skin before winter sets in.

My wife, Lisa, and I aren't much different. We can tell when it's starting to cool, and we better stock up on firewood if we want to stay warm throughout the winter. But if we are in a cubicle under fluorescent lights all day, or in a conditioned limousine with different heat settings on each of the seats, fans that blow this way or that, music playing so we can't even hear what's outside, we become so detached that the world becomes objectified, no longer living. Then we're looking at it through dead eyes. We can't get satisfaction and we can't scratch the itch and we try to make up for it by over gratifying. But it's never enough.

LW: What to do? Again, my mind goes to remembering when I was a child and unsettled, I wanted my mom, I wanted my grandmother—the nurturers. The Earth is our nurturer. She's the ultimate nurturer you know. Get muddy, get in the dirt, hug a tree, feel the pulse, feel her pulse, you know, hug her, love her, and she will love you back. It's very simple. Sometimes, when we have people come and they're in distress, before we even sit and have conversations with them about what they're feeling or going through in their heart or mind and emotions, we have them take a walk, we have them sit on the Earth. Sit amongst the trees, get their hands in the dirt, their feet, get grounded and balanced. Listen to your heartbeat in connection with the other heartbeats around you.

WW: And the whole tree hugger thing, because it became another thing… You know to hug a tree; you don't actually have to hug it. Just put your hand on it. You could put both hands! You don't have to make it like you're comforting a living person.

LW: But you can hug a tree, and it will comfort you!

WW: But really what we're saying is connect, and you can connect to the spirit of a tree by just placing your hand on it and focusing on that. If you spend any time there what you would notice is that the tree isn't just a tree; it's actually a civilization. The ants are going up there, and so are the squirrels, and the birds are nesting, and some of them eat what this tree produces. What you start to see is a whole lifecycle in this one tree, a microcosm of the Earth, and when you connect to it, that's when you would come to these other kinds of awareness—that there's more here than a tree, there's actually a community living here. So, if I were to cut down this tree, I'm not merely killing the tree, I'm also denying the validity and the existence of every other organism that relies on this tree. If you do that too much, it's the opposite of making friends, and pretty soon those friends are the opposite of friends in terms of how they receive and perceive you.

We know that the people that lived here before us were disrespectful. They committed what by Native standards would be heinous crimes against Nature, acts of cruelty actually. And we sit here, and we look at these trees that tower above us and we go "Wow, they watched all of that happen!" They watched the time before and now for the past two decades they've watched us. You know, when we started doing brush clearance it was in the spirit of protecting life here because wildfires had become so unnaturally imbalanced. When the trees actually saw us using power tools to remove brush that if ignited would be a threat to them, when they first saw us going to town on this brush clearance, they got nervous, because those two leggeds are sometimes indiscriminate about our devastation. But they soon noticed that we were working around them and that we were clearing the low brush that could endanger them and that we were creating a barrier to protect them and all of the life that they sustain. Now they smile. Now they get it.

Yes you really can make friends with trees, you can make friends with birds, and they'll remember you, you really can. There are certain species like the yellow jackets I mentioned earlier. Well, you know, sometimes they start to drown in the water that we put out for them so we'll put a stick or something they could climb onto to crawl out. But they can't always get to the stick either, so, sometimes we go over, and we've actually saved them. We make friends with that one. If you make friends with one yellow jacket—they have a very unique consciousness that they share. As soon as you make friends with one of them, somehow they all know. Start anywhere, make friends with that one.

LW: We have our bee watering dish with all the stones in it. They're very intelligent.

MK: I like that. They have this shared consciousness, and it's instant, right?

WW: Virtually. Yes, in other words it's yours for the asking. No matter how many years you've spent not connected, you could reclaim that connection in an instant and it will respond immediately, and the reciprocity comes from it, which then stimulates your reciprocity, and it becomes a perpetual motion, a feeding loop that keeps it all going. Which is why we say move in this direction (clockwise) not this direction (anticlockwise).

Which feedback loop do you want? The one that leads towards death or the one that reinforces life. Well, the design of the universe hasn't changed, it's only the two leggeds that have. And the natural world didn't mess itself up; we messed it up. That means that the condition that we're in is unnatural. Well, no wonder a lot of people seem upset—they're living unnaturally. But they can't put their finger on the button as to why, so they blame the last thing that upset them, or they blame a group of people. Or they find someone else to blame or their parents or kids. We find all kinds of things to blame instead of taking responsibility that we are the ones, each of us who sources the feedback loop that we're connected to. I like what I heard someone say recently. You can ask the Creator for anything, but if you ask the Creator for love the Creator's answer is going to be, "Oh, that has to come from you. It's already everywhere, so if you're not experiencing it, it's because you're not sourcing it." So that's the feedback loop that you created for yourself, that you keep manifesting; and every day you prove that your opinion was right, and your conclusions are right, and you're convinced of a delusion. A self-fulfilling prophetic delusion. You can prove it again tomorrow—see! Look what happened now!

MK: That's just wonderful, what you're sharing. Is there anything else you want to say?

WW: I'll speak to whoever is going to be reading this conversation and I sincerely say to you: We're strangers you and I but my prayers are with you. My heart is with you. You're not alone and we're on the same team. We have a great deal of ignorance and resistance to overcome and it's really hard to tell where this is all heading but wherever it goes don't all of us want to be on the right side of history, as the recent phrase goes? Whatever is going to happen next, don't we want to say we did everything we could realistically to try to avert the disaster because we care about the ones that haven't even been born yet and how their lives are going to start, develop, and end. I look at my seven-year-old granddaughter and I shudder. It troubles me to imagine how much worse things could be in just a few short years leading up to her adult life. So, do it for the kids. Do it for the next generation.

Lisa and I are not young anymore. You know the less future you have, the less there is to worry about. Our concerns are profound and they're not so much for ourselves. We really are worried about the people we care about, the ones who have to live in this. That's all I have to say. I appreciate the teamwork and I know that all of us having this conversation are all over the planet right now and we have to rely upon our combined efforts. We're going to find the miracle, because we're going to keep our focus on it. Good luck.

LW: Soon hopefully. I think the two leggeds need to be more like bees and become that mind that has this focus, this primary focus on the well-being and the healthy soil in which to grow our food, clean air, clean water, these basic things that life needs to survive and for the next generations to come because they are counting on us. So, thank you each one.

MK: Thank you both so much for sharing your wisdom.

WW: Wisdom is such a big word. You know as soon as you drive away, we don't feel wise.

Index

For Product Safety Concerns and Information please contact our EU
representative GPSR@taylorandfrancis.com
Taylor & Francis Verlag GmbH, Kaufingerstraße 24, 80331 München, Germany

www.ingramcontent.com/pod-product-compliance
Lightning Source LLC
Chambersburg PA
CBHW050636280326
41932CB00015B/2662

9 781032 599496